ON HISTORICIZING EPISTEMOLOGY

Cultural Memory
   in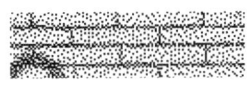
   the
   Present

Mieke Bal and Hent de Vries, Editors

ON HISTORICIZING EPISTEMOLOGY

An Essay

Hans-Jörg Rheinberger

Translated by David Fernbach

STANFORD UNIVERSITY PRESS
STANFORD, CALIFORNIA

Stanford University Press
Stanford, California

English translation ©2010 by the Board of Trustees of the Leland Stanford Junior University. All rights reserved.

On Historicizing Epistemology was originally published in German under the title *Historische Epistemologie zur Einführung* ©2007, Junius Verlag GmbH.

No part of this book may be reproduced or transmitted in any form or by any means, electronic or mechanical, including photocopying and recording, or in any information storage or retrieval system without the prior written permission of Stanford University Press.

Printed in the United States of America on acid-free, archival-quality paper

Library of Congress Cataloging-in-Publication Data

Rheinberger, Hans-Jörg.
 [Historische Epistemologie zur Einführung. English]
 On historicizing epistemology : an essay / Hans-Jörg Rheinberger ; [translated by] David Fernbach.
   p. cm.--(Cultural memory in the present)
 Includes bibliographical references and index.
   ISBN 978-0-8047-6288-5 (cloth : alk. paper)--ISBN 978-0-8047-6289-2 (pbk. : alk. paper)
 1. Knowledge, Theory of--History. 2. Science--Philosophy--History. I. Title. II. Series: Cultural memory in the present.
 Q175.32.K45R48513 2010
 121--dc22   2009046175

Typeset by Bruce Lundquist in 11/13.5 Adobe Garamond

Contents

Preface	ix
Introduction	1
1. Fin de Siècle	5
2. Between the Wars—I	19
3. Between the Wars—II	35
4. After 1945	51
5. The 1960s in France	65
6. Recent Developments	79
Conclusion	89
Notes	93
Bibliography	101
Index of Names	107

Preface

This essay is based on the readings of two seminars held in the winter semesters of 2005–6 and 2006–7 at the Technical University of Berlin. A first draft of the essay was written in the spring of 2006. An invitation of Rüdiger Campe from the Johns Hopkins University, Baltimore, allowed me to finish it in the spring of 2007. The original impulse for this book, a number of years ago, I owe to Jean-Paul Gaudillière. My thanks go to Henning Schmidgen for his critical reading of the first manuscript; to the translator, David Fernbach; and to the copy editor, Andrew Frisardi, of Stanford University Press.

ON HISTORICIZING EPISTEMOLOGY

Introduction

After the nineteenth century had seen a new empiricism in the philosophy of science, fed by the rise of the experimental sciences, the end of that century brought a particular kind of crisis—a crisis of reflection on scientific knowledge—without an immediate solution in sight, or even a generally accepted alternative to the century's legacy. Positivism, in the wake of Auguste Comte in France and the followers of Ernst Mach in the German-speaking countries, was merely the beginning of this turn, the first symptom of the crisis, as it were. Only gradually, in the course of the twentieth century, did a broadly articulated new reflection on science develop. It was fueled by various national traditions and contemporary scientific developments, and it began to historicize epistemology in various ways.

As a result, the contexts of discovery and justification, so neatly separated in between, were joined again. The idea of science as a process replaced the obligatory view of science as a system. *One single* science gave way to many sciences, not reducible to one another. This movement cannot be understood simply as something internal to philosophy or the theory of science; it must be seen in the broader perspective of a dynamics that took hold of the development of the sciences in their entirety, a process which in turn has to be placed within the social and cultural context of the twentieth century as a whole. The premise of the present essay is that the historicization of epistemology represents a decisive moment in the transformation of twentieth-century philosophy of science.

The survey that follows will present a number of authors and schools of thought, all of which played a part in this overarching movement of historicization. I will not attempt to be all-inclusive but will proceed, rather, by way of selected examples. I also will not seek to conceal personal idiosyncrasies. The order of the chapters is largely chronological, as this is how characteristic shifts can best be shown. Chapter 1 looks at the final quarter of the nineteenth century and the period leading up to the First World War. An initial role here, which must not be underestimated, was played in Germany by the famous and much discussed *ignorabimus* speech of the Berlin physiologist Emil Du Bois-Reymond in 1872. For German-speaking countries, the positivism of the Viennese physicist Ernst Mach, who rejected any kind of metaphysics, shall be compared with the conventionalist views represented in late nineteenth-century France by writers such as Émile Boutroux, from a philosophical perspective, and Henri Poincaré, from a physicalist one.[1] In Chapter 2 I discuss the 1920s, a decade that saw the first works of the Polish immunologist Ludwik Fleck and the French epistemologist Gaston Bachelard. Chapter 3 deals with the period around the Second World War. Karl Popper, Edmund Husserl, Martin Heidegger, and Ernst Cassirer all exerted a major influence in the process here considered, each in their particular way. Chapter 4 discusses the first two decades after the war, focusing on such varied figures as Alexandre Koyré, Thomas Kuhn, Stephen Toulmin, and Paul Feyerabend. Chapter 5 revolves around the poststructuralist turn of the 1960s. Its actors include Georges Canguilhem (in the tradition of Bachelard), Louis Althusser, and Michel Foucault (in turn in the tradition of Canguilhem), as well as Jacques Derrida, whose method of deconstruction took its starting point from an engagement with the late writings of Husserl. Chapter 6, finally, deals with the "practical turn" in the philosophy and history of the sciences as well as in science studies, which was also an anthropological turn represented here by Ian Hacking for the English-speaking world, and by Bruno Latour for France.

My use of the term *epistemology* requires a brief explanation. I do not use it as a synonym for a theory of knowledge (*Erkenntnis*) that inquires into what it is that makes knowledge (*Wissen*) scientific, as was characteristic of the classical tradition, especially in English-speaking countries. Rather, the concept is used here, following the French practice, for reflecting on the historical conditions *under* which, and the means *with* which, things are made into objects of knowledge. It focuses thus on the process of generating scientific knowledge and the ways in which it is initiated and maintained. If I am right, the turn from the nineteenth to the twentieth century marked a pivotal point, at which theory of knowledge in the received sense started to be transformed into epistemology in the sense in which I use the term here. This shift also marked a transformation of the problem situation. A reflection on the relationship between concept and object from the point of view of the knowing subject was gradually replaced by a reflection of the relationship between object and concept that started from the object to be known. This shift in the problem constellation is at the same time both at the core of epistemology and the point of departure for its historicization. Not by chance, an epistemology and history of experimentation crystallized conjointly. The question now was no longer how knowing subjects might attain an undisguised view of their objects, rather the question was what conditions had to be created for objects to be made into objects of empirical knowledge under historically variable conditions.

This change went with another shift of interest in the theory of knowledge. The previous orientation of finding and presenting the correct scientific method, which would be obligatory in all possible contexts, was replaced by a detailed interest in what scientists actually do in pursuit of their specific research. This gave rise to the question of whether scientists' actions, instead of following a timeless logic, were themselves subject to a historical development whose temporal course could be followed and whose particular conditions had to be ascertained. Historicization of epistemology thus also means subjecting the theory of knowledge to an empirical-historical regime, grasping its object as itself historically variable, not based in some transcendental presupposition or a priori norm.

At least to start with, a considerable part of the work of reflection that produced this turn was conducted within the sciences and by scientists themselves, rather than arising from the debates and trench warfare of academic philosophy. Thus the present investigation will also show how the process of historicization to which epistemology was subjected in the twentieth century was closely connected with the development of the sciences in this period.

In parallel with the historicization of the philosophy of science, a process unfolded that can be described as the epistemologization of the history of science. Both movements, which are to be combined under the concept of historical epistemology, give the resulting history its robustness and strength. In this connection, two events stand out above all. The first is the supersession of physics in its classical form. Connected with this, the question of scientific revolutions became unavoidable. The second is the fact, which became ever clearer, that all the sciences cannot be gathered under the same roof. This second point—and with it, the growing acceptance that it does no damage to the dynamic of the sciences if they cannot be unified, but that their plural constitution seems rather to be part of their irresistible modern drive—has developed over time perhaps still greater force. Let us now see how this development came about, and what its main lines of development were, by way of a close reading of a number of key texts.

1

Fin de Siècle

The ideal explanation of nature, as it was articulated in an increasingly radical fashion after the brief interlude of Romantic nature study in the nineteenth century, was a mechanical one. The view was repeatedly expressed that natural science ultimately aimed at reducing all phenomena down to the movements of their smallest parts and to the forces acting between them. Where this could not yet be attained, as in the case of the fundamental phenomena of life, it was hoped nonetheless that this goal would eventually be achieved with more refined methods. In certain fields, it indeed seemed that research into the depths of matter had already reached this point. Following, as its methodology, the secure guiding threads of induction, the historical course of scientific knowledge presented itself as essentially cumulative.

To the electro-physiologist Emil Du Bois-Reymond, working in Berlin, it even seemed that "the historical course of the inductive sciences is in general almost the same as the course of induction itself." In the ideal case, therefore, it was not just research and its representation that coincided, but also the method of obtaining knowledge and the actual historical process of research. Aside from the "accidents of the business of discovery,"[1] the history of a science was seen as identical with the inductive process out of which it had differentiated. For the Berlin physiologist, it was consistent with this to envisage an essentially didactic role for the history of science, a view that was echoed two decades later by Pierre Duhem: "The legitimate, sure and fruitful method of preparing a student to receive a physical hypothesis is the historical method."² In contrast to Du Bois-Reymond, however, Duhem did not share the view that a system of hypotheses could be obtained from experience by pure induction; for him, the historical method should provide a presentation of those "destinies" which had led to its introduction and enforcement. Historical representation, in Duhem's view, thus had a fundamental lesson to teach about how knowledge is attained. Neither Du Bois-Reymond nor Duhem, however, gave special treatment to this history and its methodical preconditions; the sequence of historical events remained largely unproblematical in its narrative structure.

Du Bois-Reymond gave another speech in 1872, the same year as his above-mentioned speech on the history of science at the Berlin Academy of Sciences. At the annual conference of German scientists and medical doctors in Leipzig, under the title "The Limits of the Knowledge of Nature," he drew a particular conclusion from his presentation of the progress of knowledge. Here he saw mechanical knowledge of nature—even, or indeed especially, on the assumption that it would eventually be complete—as confronting two barriers, with no way in sight as to how these could be overcome. On the one hand, mechanical knowledge was not in a position to account for its basic concepts—matter, force, movement; it could only posit them, but not according to its own inductive rules. On the other hand, it stood powerless before the phenomena of perception and consciousness. For Du Bois-Reymond, they indeed had a material foundation, but could not be derived from mechanics as understood. The fact that there was no ultimate foundation for the basic concepts with which the mechanical paradigm of scientific knowledge operated led Du Bois-Reymond to the rather radical conclusion that mechanical explanation was basically only a surrogate account, one that he liked to call an "extremely useful fiction."³

Here we see how the principle of a mechanical explanation of nature, when applied as it were to itself, collapses into agnosticism. This is the starting point of a line of thought that near the end of the nineteenth century carved out an increasingly large niche for itself under the general name of conventionalism, especially in physics and its philosophical reflection, to which I shall return. At all events, Du Bois-Reymond, with the *"ignorabimus"* at the end of his speech, provided the motto for a protracted debate, still echoing half a century later in the program of the

Vienna Circle. The focus of scientific policy, however, had changed in the meantime. In the late 1920s, against an increasing antiscience movement, members of the Vienna Circle were once more struggling for a unified science with no internal borders, if only this endeavor could be freed from the dead weight of metaphysical pseudoproblems.

Du Bois-Reymond, by positing the nonreducibility of consciousness to matter, had taken a step away from the narrow path of mechanical thinking; on this basis a dichotomy of knowledge in the form of a firm distinction between natural and human sciences could increasingly take shape. This dual distribution of competence formed the key point of attack for contemporary critics who saw Du Bois-Reymond as betraying the mechanical worldview. He met with violent, ontologically motivated resistance from the mechanically minded monist side.

Ernst Mach similarly referred both to the scope and the limits of Du Bois-Reymond's act of despair when he remarked almost a quarter of a century later:

Dubois-Reymond's recognition of the insolubility of his problem was an immense step in advance; this recognition removed a weight from many men's minds, as is shown by the success of his work, a success which is otherwise scarcely intelligible. He did not, indeed, take the further important step of seeing that the recognition of a problem as insoluble in principle, must depend on a mistaken way of stating the question. For he too, like countless others, took the instruments of a special science to be the actual world.[4]

Not to confuse the instruments of a special science and the "actual world"—the possible relativization of knowledge claims that this warning indicates shall be pursued below in the extension and in the concrete forms this relativization assumed near the end of the nineteenth century.

First of all, we must glance at how Mach himself attempted to escape the problematique that he considered as confused. His historical and critical *Science of Mechanics* can serve here as a point of departure. Mach represented research in a light that deserves closer consideration. In contrast with Du Bois-Reymond, he saw scientific knowledge less as aporetic than as a project that was on principle impossible to bring to an end. In a distant echo of Auguste Comte, Mach distinguished three epochs. The first was the animistic mythology of the old religions; the sixteenth and seventeenth centuries then saw the rise of a mechanical cosmology, still in the context of an underlying theological foundation, which in the course of the eighteenth century freed itself from religion and became the "projected world outlook of the Encyclopedists," as a mechanistic mythology. Finally, in the late nineteenth century, this mythology was overcome in a more considerate age, though it still remained to be seen how it would develop: "The direction in which this enlightenment is to be looked for, as the result of long and painstaking research, can of course only be surmised. To *anticipate* the result, or even to attempt to introduce it into any scientific investigation of today, would be mythology, not science."[5] Mach claims here that what he distinguishes as science from mythology—in the foreword to the first (1883) edition of his *Science of Mechanics* he characterized his own position as an "anti-metaphysical tendency"—is essentially determined by the lack of closure, and above all by the unpredictability of future development: on principle, we cannot see how things will turn out in the future. On the same page as cited above, Mach describes the highest form of philosophy to which a natural scientist can gain access as "*toleration* of an incomplete conception of the world and the preference for it rather than an apparently perfect, but inadequate conception." He certainly also considered his own work as a moment in an overarching historical process. He had not completed his science, but was a participant who had opened new prospects for it. On the one hand, he maintained, it was useful to make the systematic petrifications of present-day knowledge visible rather than to see these petrifications as something fixed and given once and forever. On the other hand, historical consciousness also made it possible to seek new and previously untrodden paths, "in so far as what exists is in part precisely experienced as *conventional* and *accidental*."[6] For Mach, the development of mechanics could have proceeded quite differently, and its present form was due to a contingent chain of historical circumstances. A theory of the history of science, however—unless his own three-stage sequence is misinterpreted in this light—had as little place in Mach's philosophy as in that of Du Bois-Reymond. It would arise only upon reflection on the break that positions such as Mach's introduced into nineteenth century's optimistic view of scientific progress.

In a broader historical perspective, Mach argued, the development of the sciences was bound up with the developing division of labor in society. This division created for the first time, indeed as its precondition, the need for an efficient transmission of existing knowledge in particular areas.

The social origin of science out of transmission, which for Mach was ultimately based in the biological-organic nature of man, lay at the root of his view that the nature of science really consisted in nothing more than an "economy of thought." One could state in Mach's sense that thought and its economy were ultimately the fruits of an economy of social life as this became increasingly differentiated. Concepts replace experiences that others have made and that do not have to be constantly repeated in their traditional context—a point that we shall return to, from a slightly different slant, with Edmund Husserl. Concepts are thus abstractions that can stand in for experiences at particular points. They are symbols for complexes of perceptions that show a certain stability.

Here Mach takes a definite step forward in relation to his colleague Hermann von Helmholtz. Helmholtz, with his own sign theory of perception, proceeded from the assertion that perceptions are to be understood as nerve signs for external stimuli that through experience are placed in a meaningful connection; only in this way do they become perceptions in the strict sense. Sensation is a physiological process, perception must be learned. Mach sharpened this argument almost to the point of reversing it, maintaining: "Sensations are not signs of things; but, on the contrary, a thing is a thought-symbol for a compound sensation of relative fixedness."[7] In the last analysis, it is not that we recognize unchangeable things in the external world, but rather that they are, in their thing-ness, the result of an effort of abstraction. There is no such thing in nature as cause and effect—concepts that Mach calls "things of thought, having an economical office"[8]—but merely concrete connections in which similarly concrete individual subjects with their perceptive capacity always already find themselves.

This does not mean, however, that the conventional and the accidental in the development of knowledge are lacking in rules. Indeed, if Mach locates the genesis of these "things of thought" with their "economical office" in the biological-organic, this is not the case with their enforcement. These things are rather subjected to a historical regime of economy which prevents their random development in any direction. Mach thus describes science in general as "the least possible expenditure of thought."[9] The basic concepts that a science such as mechanics deploys are the result of a principle of least expenditure of thought: the way they represent the mechanical facts of experience can be undercut at any time. What is still more important: "The science of mechanics does not comprise the foundations, no, nor even a part of the world, but only an *aspect* of it."[10] Mechanics no longer forms the center of the natural-scientific explanation of the world on account of its ontologically privileged object, as was still the case with Du Bois-Reymond. Its only justification is that of a proven economy of thought. Every such economy, however, grasps only one *aspect* of the world; it is an abstraction from *one* particular perspective. And the perspectives from which the world can be grasped are on principle limitless, with none of them being specially privileged. There are other sciences that abstract according to other principles, and these can have the same validity as mechanics.

. . .

Henri Poincaré in turn was one of the most prominent representatives of a theory of science that went under the already mentioned name of conventionalism. Before we compare his line of argument with that of Mach, however, we should briefly consider Émile Boutroux—one of the most influential philosophers of science in France in the late nineteenth century. His work on the sciences broaches themes that we shall discuss in their later elaborations, while also linking up with Du Bois-Reymond and Mach. Unlike his German peers, Boutroux was a trained philosopher, though he concerned himself intensively with the sciences of his time. He can be seen as one of the fathers of a rapprochement between philosophy and the natural sciences in France, which later was to flow into a special form of historical epistemology. His 1874 dissertation *The Contingency of the Laws of Nature* articulates all the themes that are relevant to our present concern.

Contingency was the key word for Boutroux, with which he sought to break up the determinism of classical mechanics. His line of argument requires closer examination. According to Boutroux, we obtain our information about the world solely from the empirical sciences, which approach their objects by way of experiment: "All experimental finding is reduced, in the end, to confining within as close limits as possible the value of the measurable element of the phenomena." Yet "we never reach the exact point at which the phenomenon really begins and ends."[11] In between there is a space of indeterminacy, which the determinist argues away by "interpret[ing] literally the principle by which any particular phenomenon is connected with any other particular phenomenon."[12] This step, however, has no justification. A de facto space of indeterminacy exists at the heart of quantitative science, and nothing justifies us to make assertions about a realm that already escapes the experimental means of our assessment in the context of what Boutroux describes as the "static sciences." This holds all the more so for the "dynamic sciences"—the "sciences of being"—and naturally of course also for the historical sciences. It would be wrong, says Boutroux in his second major treatment of the concept of natural law, based on his lectures at the Sorbonne some twenty years later, "to say that mechanics, of itself alone, constitutes the entire science of the real. For in the present state of our knowledge, science is not one, it is multiple. Science, regarded as including all the sciences, is but an abstraction."[13]

Contingency, however, not only defines a free space of indeterminacy, it equally marks the condition of possibility for development and history. This holds a fortiori for that human activity that is the expression of human particularity par excellence—scientific activity and the genuine historical development by which it is specified. In the end, research itself becomes the highest expression of the principle of contingency:

But if it is legitimate to set up dynamic sciences alongside of and above the static sciences; if objective science actually consists of these higher sciences, then the doctrine of contingency is conformable to the conditions of science. The only thing is that this doctrine imposes observation and experiment as the ever indispensable method of the dynamic sciences, the sciences of being. If indeed, along with a principle of conservation there is also one of contingent change, the abandonment of experiment is always dangerous and illegitimate. No longer is experiment a confused thought, a chronological starting-point of separate thought; no longer is it even the totality of the data amongst which induction discerns law, and which, once thus summed up in a general formula, render new observations ineffectual: it is the eternal source and rule of science, in so far as this latter would know things in truly objective fashion, i.e. in their history as well as in their nature, which, after all, is but one of their states. According to the doctrine of contingency, it is erroneous and chimerical to attempt to reduce history to a simple deduction.[14]

Objectivity, in the last analysis, thus becomes a historical task. It is not some kind of essence of things that deserves our highest attention, but rather their history, and this naturally holds above all also for scientific research as a permanent process of contingent variations. Yet Boutroux does not go so far as to include philosophy itself as a reflection of science in this process. His philosophy of science remains as abstract as the concept of *science* in the singular which Boutroux so decisively criticized.

Boutroux distinguished two kinds of things as the result of scientific activity: those that are "more akin to mathematical conjunction and imply considerable elaboration and purification of concepts," and those that are "nearer to observation and induction, pure and simple."[15] The two are committed to different ideals, and hence can be brought together only in the final analysis. There is one point however that they have in common: they always fall short of these ideals, for what we describe as natural laws is at bottom "the sum total of the methods we have discovered for adapting things to the mind."[16]

. . .

In his influential *Introduction to the Human Sciences* of 1883, Wilhelm Dilthey brought the attitude toward scientific concept formation so characteristic of the fin de siècle onto the following curt common denominator: "Most leading natural scientists consider such concepts as force, atom, and molecule a system of auxiliary constructions which help us develop determinations of the given into a system which is clear to understand and useful for life. And this corresponds to the facts of the case."[17] Thus these concepts are not "entities with a higher rank than an individual object," but rather "historical products of the logical mind struggling with objects," results of the "development of a partial system given in nature as an abstract device for knowing and exploiting nature."[18] Understanding this development as a historical process consists in drawing a distinction, if not opposition "between the historical consciousness of the present and any kind of metaphysics as a scientific world outlook." The essential thing for Dilthey, the signature of the time that also embraces the natural sciences, is therefore "insight into the development itself," which by no means, however, excludes the goal of "objectively knowing the connection of reality."[19]

This also ties up with the moderate and restrained conventionalism of Boutroux's brother-in-law Henri Poincaré. "It is clear," he wrote in 1902 in his first published book, which also cited Mach's preface to the fifth edition of his *History of Mechanics*, "that any fact can be generalized in an infinite number of ways, and it is a question of choice. The choice can only be guided by considerations of simplicity."[20] Mach's principle of economy reappears here as a principle of simplicity. Yet Poincaré defends himself against a nominalist intensification of his position: "Many people have exaggerated the role of convention in science; they have even gone so far as to say that law, that scientific fact itself, was created by the scientists. This is going much too far in the direction of nominalism. No, scientific laws are not artificial creations; we have no reason to regard them as accidental, though it be impossible to prove they are not."[21] What protects us from an overly radical conventionalism is industry and its technological systems: they successfully translate scientific thought. "Fortunately, science leads on to application, and this puts paid to the skeptics."[22] Poincaré pleads for a realism of "relations" that are grasped in approximate laws. Particular relations—especially if they have acquired the character of "simplicity"—are transformed into principles on grounds of "convenience," and thus escape further control by experience. The language in which they are expressed changes, however, and the objects between which these relations hold remain hypostasized and subject to a continuous change of meaning; they are ultimately exchangeable.

At the first blush it seems to us that the theories last only a day and that ruins upon ruins accumulate. Today the theories are born, to-morrow they are the fashion, the day after to-morrow they are classic, the fourth day they are superannuated, and the fifth they are forgotten. But if we look more closely, we see that what thus succumb are the theories, properly so called, those which pretend to teach us what things are. But there is in them something which usually survives. If one of them has taught us a true relation, this relation is definitely acquired, and it will be found again under a new disguise in the other theories which will successively come to reign in place of the old.[23]

These stable relations between things may well be the product of our minds, but in the last analysis they are obligatory for all thinking people, and to this extent are also for Poincaré the point at which it remains possible to speak of objectivity.

An interesting comparison can be made here, which Poincaré draws in respect to the "progress of science." In *The Value of Science*, he compares the history of the sciences with the evolutionary emergence of animal forms.

These experience multifarious transformations in the course of time, which lead them to appear so far removed from one another that the uneducated gaze may no longer recognize any affinity between them. The practiced eye, however, will be able to see in today's patterns of thought the traces of their origin. Despite the considerable extent of these transformations, invariants can be distinguished that underlie the building plans of knowledge. These are the "relations" that Poincaré privileges. Whether such connections will last is a question that cannot be answered a priori.

What Dilthey in his remarks on "historical consciousness and world outlooks" demands of a philosophy for a postmetaphysical age—that it take itself as an object in the context of the dissolution of all metaphysical orientation of man into reality as a "human-historical fact,"[24] meaning above all its own presentation as a *historical* fact and a historical problem—applies in the same measure also to the procedures of the natural sciences. It is no accident that this new form of scientific self-reflection should appear at more or less the same time as the disappearance of belief in the possibility of a unique and all-embracing metaphysical system. This scientific self-reflection initially took the form of epistemological considerations that, against the background of rationalist or empiricist traditions, began to ask whether the business of acquiring scientific knowledge could be subsumed under a single method. In the first phase, this argument only marginally concerned the history of science; moreover, it is striking how many authors emerged from a scientific background. Only a minority came from academic philosophy concerned with science, which at this time, in Germany at least, ramified into the many variants of neo-Kantian theories of knowledge. This reflection was, however, confronted with a certain dilemma: in light of the contemporary pluralism of philosophies and world outlooks, the natural sciences had on the one hand to continue to present themselves as a relatively united undertaking, which had been far from unsuccessful in the nineteenth century in outrunning religious authority, although on the other hand, breaches and irreducible diversities became apparent in their own camp.

. . .

The Viennese sociologist, economist, and philosopher Otto Neurath, whose small oeuvre on the history of science is less well known than it deserves to be, brought together a series of formerly scattered considerations in a programmatic essay of 1915—"On the Foundations of the History of Optics," published in the *Archiv für die Geschichte der Naturwissenschaften und der Technik*—and explained with the example of optics how the outlines of a history of science could appear, superseding mere chronology and psychology with an epistemological grasp on their material. For Neurath, the history of science became a constructive enterprise, in which both Dilthey's diagnosis of epochs and the view that mathematicians like Poincaré or physicists like Mach had presented of the structure and development of physical theories had a part to play. Neurath undertook here an unconstrained first attempt to relieve the history of the sciences of its merely descriptive character and to pursue a kind of historical analysis. The object was in this case less to provide the philosophy of science with historical coordinates than to provide the history of science with an epistemological arsenal.

Neurath opened his essay with the following statement:

History of science, if seen as more than a mere chronicle of findings and biographies, is a young discipline. It can aim much higher: like the history of any field of enquiry, it may try to shed light on the psychology of the enquirer; besides, it may exhibit the logical structure of theories, and from it derive how they may develop. To follow how such possibilities happen to be realized by this or that enquirer is an especially engaging task.[25]

Neurath considered such logical analysis a precondition for the reconstruction of developmental options against which the course of history could be plotted. In a second work from the same year, "Classification of Systems of Hypotheses," he described what had so far been wanting as a "special technique for the analysis of trains of ideas."[26] Neurath made a start on implementing this vision in the two publications just mentioned, with respect to a series of basic ideas in the history of optics and the changing combination of these ideas.

The establishment of such a technique, on which Neurath believed it was possible to agree, could in his view also help with a further lack—that of the "continual cooperation of scholars in the field of the history of science."[27] Historians of science still had to learn to work together on the history of scientific disciplines, as for example chemists did in chemistry. They lacked "a routine, partly practical and partly theoretical," the tools of their trade, as it were, such as anatomists might have in the procedure for dissecting a corpse.[28] If they wanted to work together successfully, then historians of physics, for example, had to "arrange the views of physicists into groups in the same way as botanists, the plants, or chemists, the compounds."[29] Neurath sought to make clear, on the example of optics, that the consequences of this move were by no means trivial. The history of optics, according to Neurath, had been written for the most part as the history of a dualism between the particle theory and the wave theory of light. It would have assumed a quite different form "if physicists had been divided into 'periodists' and 'non-periodists.'"[30]

Neurath viewed this technique of analyzing the course of ideas, however, only as a precondition for the successful pursuit of history of science. The development of trains of ideas in the natural sciences, whether these were at times more strongly marked by change in empirical data or change in interpretation—and in such a history, according to Neurath, the two phases regularly alternated—had to be seen in the wider context of a history of ideas about the world: "In order to really do justice to trains of ideas in physics, these have to be articulated with the idea of the world as a whole." For all scientists participate, even in forms that may be very vague and perhaps unconscious, in the pictures of their contemporary world, which "prop up" their work in one way or another.[31] For Neurath, then, it was not accidental that, up to today, "genuine contributions have come only from those historians of science who were familiar with the philosophy of their time."[32] Among those of his own time, of course, Neurath placed Ernst Mach first.

Two things, in his view, characterized a history of science that made theoretical claims. On the one hand it had to uncover the key elements for a historical reconstruction. Only this created the analytical distance from the historical object that was needed. The historian could no longer simply take over traditional categorizations. In seeking out optional spaces of this kind, it was less important to know in detail what individual researchers had actually thought. No psychographs were needed; the point was rather to expose the conditions in which their scientific system could take shape, to "stat[e], for each theory, the essential parts of experience."[33] This aspect, however, which is more a function of classification, was overlaid by another interpretation, one that could be equally decisive for understanding a particular historical course: "We must try to see clearly how far a physical theory hinges on the images used, and how far on those features that actually carry the argument. Perhaps we cannot grasp some developments unless we consider the images and pictures; in other cases, we must rely on what governs the mathematical treatment of phenomena; or, maybe, both ways of looking at it are steps."[34]

A large number of "leading images," as Neurath called them, may well owe their long persistence in the history of science less to the fact that they give a comprehensive account of the ramifications of the phenomenon being measured than to the fact that they are simply very convenient. With this—to cite Neurath again, who presents this argument in the context of the philosophical debate just referred to—a bridge would be built "with the type of contemporary philosophy held by Poincaré who said, concerning emission and wave hypotheses, that their role was secondary, and that they were retained merely for convenience of clear exposition."[35]

Neurath's guiding threads for a theoretically fruitful history of science can thus be arranged in three circles. The center is the analysis of those elements of a system of hypotheses that lend themselves to calculation. At the same time, these elements always reveal something about the neglect of certain phenomena and the privileging of others. In Neurath's words, they involve not only "an indication of connection"—Poincaré's "relations"—but also the selection of facts.[36] The systems, for their part, are embedded in "leading images" that endow an area of experience on which a science develops with what Neurath at one point calls a "fuzzy margin." These pictorial worlds, in turn, in which analogies play a central part, find their place in a still "more thoroughgoing exploration of how ideas develop," which in the last analysis must go back to the "innermost drives of the history of ideas."[37] "Our reference to a total world-view becomes a duty."[38] Neurath stresses: "As we need theories to classify things, so we need theories to classify theories."[39] He was convinced that a historical investigation into scientific theories about the world needed a metatheory of this kind based on comparison; only in this way would it develop into a rewarding and fruitful enterprise that itself deserved the name of science.

2

Between the Wars—I

This chapter will focus on two twentieth-century thinkers, each of whom in his own way tried to transform the historicization of epistemology into a comprehensive program. Though in all probability they were unknown to one another, and as far as we can tell were unfamiliar with each other's work, they demonstrate—despite their different backgrounds—a surprising number of points in common. These two are the Polish immunologist Ludwik Fleck, theorist and sociologist of science, and the French philosopher of science Gaston Bachelard. Both are likewise representatives of a time characterized by a major political as well as a major scientific event.

The major political event was the First World War, which marked a deep caesura in the conception of scientific progress characteristic of the nineteenth century. For the first time, the ambivalence of science as a means both of technological construction and of mass destruction on a large scale and with devastating consequences became clear. After 1918, the relationship between science and technology appeared very different from how it had looked before the Great War. The material conditions of the production of knowledge, which previously mattered only marginally or rhetorically to those who reflected on the development of the sciences, acquired central importance in historical consideration. The relationship between science and industry also appeared in a new light, as did that between science and the state. It was not that everything suddenly changed; but the event of the war marked one of those junctures in the historical landscape from which earlier developments are also viewed in a new perspective—implicitly or explicitly. This was true not just of Marxist tendencies in the discussion of the sciences, which in the terminology of traditional historiography of science came under the rubric of "external" or externalist historiography, and which in the 1920s and 1930s were oriented to historical studies on the social, political, and technological conditions for the production of knowledge, as represented by authors such as Henryk Grossmann and Boris Hessen. It was perhaps still more decisive, as we will see, for an "internal" approach that came to conceive of modern scientific thought fundamentally in technological terms; in its wake, the dynamic of scientific development no longer appeared only as a matter of cumulative steps, but also of revolutionary breaks.

The major scientific event was the shock that classical physics experienced, following the appearance of nonclassical geometries in the late nineteenth century, first with Albert Einstein's theory of relativity and then with the subsequent overwhelming developments of quantum physics in the course of the 1920s. How deep this shock went is clearly visible in the cases of both Fleck and Bachelard. A decisive point was that one could not avoid accepting the existence of theoretical alternatives, even in the hard sciences. It was no longer simply a matter of the gradual further development of concepts already present, but rather of the formulation of entirely new ones. A second theme, which repeatedly arose in the most varied connections, was the problem of the dependence of measurement on the observer, vigorously discussed in the context of quantum physics. The "subjective" moment, which thus came to play a role at the very heart of the natural sciences, raised anew not only the question of the relationship between the natural and the human sciences that writers such as Du Bois-Reymond or Dilthey had marked out in such a balanced way, in both its scientific and its philosophical aspects, but that of the very possibility of a particular form of objectivity. The revolution in early twentieth-century physics had a further consequence: it rekindled a debate about the feasibility of the goal of a unified science, and whether this goal might actually be unnecessary altogether.

A common feature of the two thinkers discussed in this chapter is that neither was attached to a traditional school of philosophy of science. Both Fleck and Bachelard were outsiders in the eyes of their contemporaries, belonging to no definite tradition. Although in a completely individual fashion, and from a different specialist perspective, each of them placed central importance on two things: first, the experimental and technical character of modern science, and second, its social character.

. . .

Bachelard's path led him from post-office employee, via teacher of physics and chemistry at a secondary school in his native Bar-sur-Aube, to the University of Dijon in 1930, and finally in 1940 to the professorship of history and philosophy of science at the Sorbonne, where he succeeded his former teacher Abel Rey. From his first book, *Essai su la connaissance approchée* (Essay on Approximative Knowledge), published in 1928, Bachelard's central concern remained that of characterizing what he himself called, in the title of his 1934 work, *The New Scientific Spirit*. This new spirit, which Bachelard sought and depicted in particular in physics and chemistry, escaped in his view the traditional alternative between rationalism and empiricism/realism; and perhaps, Bachelard suggested, it might be quite inappropriate to see modern scientific practice as always and everywhere the same. It was better first of all to examine precisely what went on in actual laboratories, and the procedures followed there. It was not the task of philosophers of science to dictate to scientists the conditions of possibility and the norms of their knowledge; they should rather familiarize themselves with the laboratories and workshops of science, and especially with the history of science as the epistemological laboratory par excellence. "Sooner or later, scientific thought will itself become the central object of philosophical reflection. . . . Philosophy must therefore modify its language if it is to reflect the subtlety and movement of contemporary [scientific] thought."[1] Bachelard was no longer referring here to philosophy of science alone. He was convinced that contemporary philosophy tout court had to orient itself toward an analysis of scientific thought, and seek its main thematic focus there. His key conclusion was that there is no single system that everything fits into. We have to live with this "lack of metaphysical purity." It is based in the dual character of the scientific procedure, in which experience and thought, reality and reason are related to one another in such a way that it is not possible to privilege one aspect over another, since they stand in an indissoluble reciprocal interaction. *"Experimentation must give way to argument, and argument must have recourse to experimentation. Every application is a form of transcendence."*[2]

Bachelard's concept of "realization," therefore, turns out to be a key concept for the understanding of his position that became known as historical epistemology. Modern science verifies by way of realization. This corresponds to a "technological realism," which in Bachelard's eyes is a characteristic feature of contemporary scientific procedure and thought. The question is not one of reality and its knowledge—that which *is*—but rather one of "realization," that is, what *can be*. "It is rather a realism at one remove, conceived in reaction to the usual notion of reality, as a polemic against the immediate; it consists of realized reason, reason subject to experimentation."[3]

Here Bachelard touches on another important theme, which he developed in more detail in *The Formation of the Scientific Mind*, and which has fundamentally determined the reception of his work: the notion of the "epistemological break." Scientific thought emerges from a break with everyday experience, "in the struggle with immediacy." It is counterintuitive and so is not confirmed by the evidence of direct observation. The latter rather functions as an "epistemological obstacle."[4] The constitution of modern scientific thought is mediated by instruments. An epistemological break and an epistemological obstacle are always related. They characterize not simply the transition from everyday experience to scientific experience, but they are of lasting actuality also for the further development of the sciences. Any discovery that is genuinely new proves to be a break with what was previously scientifically given; it is realized in polemical form. "Every new truth comes into being in spite of the evidence; every new experience is acquired in spite of immediate experience."[5]

The "reality of science"—*le réel scientifique*—is thus realized in a kind of spiral of supersession or transcendence. What Bachelard calls reality of a second order should not be thought of as a Kantian thing-in-itself, but rather as a context of experimental axes in which the concepts at work and thus the "real" concepts are inscribed. "The time of the adaptable patchwork hypothesis is over, and so is the time of fixation on isolated experimental curiosities. Henceforth, hypothesis is synthesis."[6] Experimental contexts are recursive arrangements in which new knowledge arises that constantly challenges to rethink the presuppositions of the method in use and calls for adjustment. "Hence there is a need for methodological innovation . . . ; theory and experiment are so clearly related that no theoretical or experimental methodology is guaranteed to retain its validity indefinitely."[7]

From the perspective of the objects that are realized in such experimental contexts, Bachelard spoke of modern science as a "phenomeno-technology." In the experiment, phenomena have to be "selected, purified, shaped by instruments; indeed, it may well be the instruments that produce the phenomenon in the first place."[8] Conversely, the phenomena produced in the experiment for their part provide the occasion to problematize the theoretical assumptions embodied in the instruments. Phenomeno-technology "takes its instruction from construction."[9] The philosopher and art historian Edgar Wind, in his habilitation treatise, *Experiment and Metaphysics*, which was contemporary with Bachelard's *New Scientific Spirit*, spoke in this respect of a circle that basically cannot be brought to a halt—not in the context of "restriction . . . to a view of the world based on instruments which have their place within this world and are therefore subject to the laws of the world," as is demanded by and peculiar to experimental science. From this restriction, it follows, "to make precise use of those instruments we must know the laws of the world to which they are subject. On the other hand, it is precisely the goal of this use to find out these laws in the first place."[10]

It is exactly this circular connection that drives the process of obtaining experimental knowledge. In this process, knowledge is not simply iterated, but is differentially driven forward. If the scientific spirit has once established itself on this path of embodiment of knowledge, it *must* keep its instruments and experiments, as well as its concepts and theories, in permanent motion and change; not because modern science consists in some search of agreement between subject and object, mind and reality, that can be precisely grasped, but because it is a "project": "In scientific thought the subject's meditation upon the object always takes the form of a project."[11] The project is a draft, and science is realized in drafting. The draft, however, only exists in the form of alternatives with possible experimental realizations. What Bachelard demands, taking over a concept from the contemporary debate over complementarity in microphysics, is precisely an "ontology of complementarity."[12] It can be addressed as a kind of epistemology of difference, a mild variant of a dialectic of contradiction, thus a theory of process or theory of development of the scientific spirit. With his position, Bachelard locates himself firmly on ontological ground. He polemicizes against both conventionalist predecessors and pragmatic positions, which he sees as capitulation in the face of the challenge to develop an adequate epistemology of the modern process of knowledge acquisition, and whose "scattered pluralism" and voluntarism he rejects in the name of that to which he claimed to adhere: a "coherent pluralism" of scientific knowledge.[13]

A second point, which deserves further attention here, is Bachelard's picture of the social constitution of science. Community comes into play at several levels. The first of these takes us to the center of Bachelard's "non-Cartesian" epistemology.[14] The modern process of knowledge acquisition must be conceived of as a fundamentally mediated one, in which the Cartesian relationship between the thinking ego and the world to be known is put in question in two respects. First, as already indicated, the production of scientific knowledge is in a fundamental sense instrumentally mediated. Subject and object do not face one another directly in the experiment, but are engaged in a process of mutual instruction. Just as the techno-phenomena that science investigates are not simply there but only realized in the experiment, and in their experimental individualization must be reconnected constantly, so the scientific spirit is only realized in and through this process. It exists only as a history of involvement in and entanglement with the phenomena that it investigates. Second, the modern scientific working process is the result of the work of many people, and must therefore also be negotiated between many subjects. Contemporary science is a collective undertaking, organized and carried out by a community. Bachelard sees "objectification" as having both a real and a social significance.[15] Here too he speaks of a process. Just as what experimental science deals with is not so much an established reality but rather material *realizations*, so objectivity is not something given but rather is produced in a process of *objectification*, the result of a double instruction—of phenomena and of minds.

A second level is given by the differentiation—not to say fragmentation—of the various fields of the experimental sciences. They form many islands or cultures of "access to an emergence,"[16] as Bachelard put it in a later text. This differentiation is grounded in the nature of experiment itself, which ultimately must always be performed on particular entities in concrete contexts. Different microrationalities are developed around these particular entities, which may be marked by quite different sets of problems and surrounding material conditions, and which each have their own epistemological awkwardness. According to Bachelard, however, philosophy of science must accept these details if it is to raise itself to the level of the sciences of its time. In *The Philosophy of No* he called for a "dispersed" or "distributed" philosophy.[17] Every significant complex of problems, every more or less fruitful experimental arrangement, every equation even, demands its own philosophical reflection. Bachelard's whole work is a plea for an epistemology of detail of this kind.

In contrast to many of his contemporaries, who complained about the process of fragmentation of the sciences into ever more specialized disciplines, and misguidedly aimed at a comprehensive scientific formation, Bachelard emphasized the productive aspect of the process. The islands of scientific rationality are not disciplines with hard and fast borders, but rather little patterns, cantons, or regions that may suddenly arise on the basis of an instrument or experiment, but whose margins are fluid and can disappear or merge together with other unities. Epistemologically, these units are highly flexible. Precisely because they are only loosely linked with other regions, they can easily experience transformations and substitutions, from which only at a later time will it become obvious as to how fruitful they are, if at all.

Bachelard approached the sciences not in terms of the structure of their academic organization but rather in terms of the structure of the research process. The social units that are formed here are not firmly circumscribed and fixed institutions, but rather small and informal communities of scientists, which at times can take the character of idiosyncratic schools. The research they conduct forms a highly mobile ensemble, which ultimately however is held together by a particularly strong nexus—the common connection with the world of technology. Just as at the very heart of the experimental process everything revolves around techno-phenomena, so the natural sciences as a whole can be socially understood only in terms of the spirit of technology, to which they are related as the field of their realization on a social scale. Technological application, which according to Bachelard is to be seen as the basic social constitution of the modern sciences, constantly drives the sciences to transcend their own limits but finally also holds them together. This last aspect Bachelard particularly emphasizes in his late epistemological work from the time shortly after the Second World War. Here he takes up and further pursues a theme that we already encountered with respect to Poincaré in Chapter 1, although there the argument of technology was mainly used to moderate conventionalist speculation.

Finally, we need to look at the particular form of historicity in Bachelard's epistemology. One of his fundamental statements is that "the scientific spirit . . . is essentially a way of rectifying knowledge." "Sitting in judgment," the scientific spirit "condemns its historic past. Its structure is its awareness of its historical errors. For science, truth is nothing other than a historical corrective to a persistent error, and experience is a corrective for common and primary illusions."[18] For this process of permanent correction, which begins with the break with "common sense," Bachelard uses the concept of "recurrence." As we have already emphasized, a constant process of reorientation takes place in scientific action itself, once it is set in motion, permanently transforming the truths of yesterday into the errors of today. Since the truth of today, however, is again only produced in this movement of negation—is constituted essentially in a polemical fashion—it also experiences this notable dual character: on the one hand it is a correction of the past, on the other a defendant in a permanent tribunal. At bottom, every scientific truth of today risks ending up as an error of the past. This makes for the particular—and what can even be called constitutive—historicity of the sciences; they are constantly driven beyond themselves and yet remain recursively related to their problem states in the sense of a historical linkage. It is precisely in this way that "the history of the sciences appears as the most irreversible of all histories."[19]

From the moment that knowledge is realized in such polemical linkages, and for the very first time "begins to have a history," the scientific mind also assumes a "variable structure,"[20] in that it pluralizes and diversifies precisely in its concentration on a particular issue. Scientific thought, accordingly, cannot be characterized as a system of propositions; it is rather a process of evolution. It finds its justification, not in the unity of a thinking ego, but rather in the historical structure of its replaceability. Non-Cartesian epistemology, according to Bachelard, is "by essence and not by accident in a constant state of crisis."[21] Every method of gaining knowledge is determined to "risk itself in new acquisition," and it obtains this right of risking itself on the basis of its "legitimization by earlier acquisition."[22] Such historical irreversibility, however, is not to be understood as grounding a necessary development in the sense of a teleology of progress. There is no "historical reason" at work, in the Hegelian sense, in the displacements of the history of science. For Bachelard, the achievements of science are and remain emergent phenomena. They have the character of events, and although they form a chain, the individual links remain historically contingent. In the words of his contemporary, the French quantum physicist Louis de Broglie, Bachelard—following on from Mach—held that "many scientific ideas of today would be different from how they are if the paths that the human mind took in pursuing them had been other."[23] Finally, it is up to historical epistemology, insofar as it takes seriously its own actualistic claim of seeking to rise to the level of its age in relation to the most recent scientific developments, that it transcends itself and understands its own knowledge production as a process of historical change. "If we pursue the ideal of modernistic tension, which I demand for the history of science, then it will be necessary to constantly rewrite and reconsider the history of the sciences."[24] Historical epistemology, in this sense, not only has to do with the historicity of the sciences, but is itself a historical enterprise.

. . .

The second thinker to be discussed in this chapter is the immunologist Ludwik Fleck. Fleck studied medicine at the University of Lwów, where in the late 1920s and early 1930s he wrote his decisive reflections on science while at the same time conducting bacteriological research. His most important monograph, *The Genesis and Development of a Scientific Fact*, was published in Basel in 1935. Following the Nazi invasion of the Soviet Union, Fleck was confined to the Lwów ghetto and eventually survived the concentration camps of Buchenwald and Auschwitz. After the war, he continued his immunological work at the Polish Academy of Sciences. Only some fifty years after they were written did his texts find a broader reception.

Fleck formulated the basic coordinates of his conception of science in an essay with the title "Zur Krise der 'Wirklichkeit'" (On the Crisis of "Reality"), published in 1929 in the German science journal *Die Naturwissenschaften*. Its title echoed an essay by the cultural politician and philosopher Kurt Riezler, which had been published in the same periodical the previous year, treating the perceived "crisis" from a double perspective. On the one hand, according to Riezler, recent scientific development had downgraded the concept of natural law to a statistical quantity, while on the other hand it had become clear that the individual sciences, instead of advancing toward a unified science, were becoming ever more divergent.

Fleck took up the buzzword of "crisis" in order to argue for a fundamentally different "theory of knowledge," as he put it, the purely logical variants having proved inadequate in his view. "If we investigate the sources of knowledge," his essay began, "we generally make the mistake of imagining them in far too simple a way."[25]

Fleck distinguished three factors, or rather three systems of factors, that played a constitutive part in our "psychology of knowledge": the "burden of tradition," the "weight of training," and the "effect of the sequential order of knowledge." More closely considered, however, all three converge on the same point, which Fleck formulated as follows: "These are social moments, and each theory of knowledge must therefore be brought into connection with the social realm and through this with that of cultural history, if it is not to end up in a serious contradiction with the history of knowledge and the daily experience of teachers and students."[26] According to Fleck, therefore, a theory of knowledge cannot proceed from the individual, that is, as an "individual matter of a symbolic 'man,'" it must rather assume from the start a transindividual, social, and cultural structure. It is not reducible to an elementary subject-object relationship. "Wherever and whenever we begin, we are always already in the midst of things."[27]

In the passage just quoted, Fleck mentions two sources for the construction of a new theory of knowledge: the history of knowledge and the daily experience of the researcher and teacher. In *The Genesis and Development of a Scientific Fact*, Fleck described a theory of knowledge not based on historical and comparative investigations as an "empty play of words," an "*epistemology of the imagination*"[28] located in the realm of speculation. With respect to the history of the concept of syphilis and the development of the Wassermann reaction as a test for the disease, he demonstrated in detail how the history of science could be used in order to shape basic categories for a theory of the sciences—such as "style of thought" and "thought collective," to which we shall return. Here we need only indicate that Fleck's starting point was a dual view of knowledge as a process. On the one hand, natural science is "a work that is never at an end, eternal like the work of a stream in forming its bed."[29] This image of a stream forming its bed, however, refers to yet another point that was central to Fleck's historical conception of knowledge. Knowledge does not proceed asymptotically toward something like an absolute reality; rather it moves away from something: it is fundamentally "development-conditioned," that is, determined by the steps it takes. What can be discovered is laid down by the successive series of preceding discoveries.[30] The internal historicity that is thus brought into play can no longer be formulated from the perspective of a knowing subject; it is rather the product of a history as process. "Like everything socially conditioned, what is known has its own independent life quite distinct from the individual, its properties, its style of time and place, and as a result also its own fate."[31]

Second, the experience of the scientist—of many scientists in all possible fields—is needed in order to obtain a robust picture of the reality of science. The "living practice" of the natural sciences must be distinguished from their "official paper form." It is differently structured, not proceeding as described in books of logic or in textbooks. The practice of the natural sciences in general "cannot be learned from any book,"[32] it can only be found in and learned from the reality of the laboratory. With this demand, Fleck—himself both observer and participant in laboratory life—largely anticipated the sociology of the laboratory that would develop only in the last decades of the twentieth century.

Like Bachelard, Fleck was also concerned with the epistemological shocks brought about by quantum physics, in particular the question of the interaction between observer and observed, as it presented itself, according to Bohr, in the observation of atomic phenomena. Fleck, who in his essay explicitly and specifically referred to Bohr, understood an interaction of this kind as constitutive of knowledge in general. "Observation, discovery, is always feeling one's way, that is, literally a reshaping of the object of knowledge."[33] Discovery, for Fleck, was a "relationship of active, living intervention, a reshaping and being reshaped, in brief a creation."[34]

Also like Bachelard, Fleck conceived the relationship of subject and object in the process of discovery no longer as analytic and contemplative but rather as synthetic and constructive, accompanied by phenomena of emergence that would become characteristic of entire "series of discoveries."[35] This synthesis and construction may well bear the personal stamp of the work of individuals, but in the end it is always indebted to a collaboration of groups of people. Fleck stressed more strongly in his early work than in his later writings how for him the way of thinking of the natural sciences appeared as deeply "democratic." It not only demands openness toward everybody and thus places knowledge under a control that is in principle public, it must also let itself be informed about something better.

Fleck even goes as far as to characterize the work of the community of natural scientists as the social role model of a democratic society: "For natural science is the art of forming a democratic reality and orienting oneself according to it—thus being reshaped by it."[36] Natural science and democratic society, for Fleck, form a kind of epochal stylistic unity, which differentiates the modern age from earlier ones. It is not that scientific thinking and action were absent at previous times; they were always present in the realm of handicraft, but not yet dominant in a style-forming sense.

The concepts of style, more precisely "thought style," and "thought collective" form a central point of reference in Fleck's works on the theory of knowledge. The two concepts therefore deserve to be looked at in somewhat more detail. They are related, and based on Fleck's conviction that science is a process carried out by groups and not the result of acts of lonely individuals. In his essay *The Genesis and Development of a Scientific Fact*, Fleck describes thought style as a concept into which "all paths toward a positive, fruitful theory of knowledge lead."[37] Thought style should not be viewed as a logical system, its characteristic is rather the "readiness for directed perception." Fleck decidedly demarcates himself here from what he considered a basic error of logicist and positivist theories of knowledge—from Mach to Carnap—that is, the assumption of an observation with no presuppositions: "psychologically . . . nonsense and logically a game."[38] The "scientific fact" is not the starting point of observation but rather the result of the production of a habit of perception. It is a matter of style of thought, and thus the result of a movement from undirected to directed vision. Directed vision, however, is always due to practice. The concept of "being experienced" thus acquires a "fundamental epistemological importance," along with "its hidden irrationality."[39]

But the style of thought that sums up the overall vision of a community of scientists is not just the formation of a convention. It rests on a material foundation. Fleck is interested in what happens "while making the effort of acquiring [a finding]."[40] It is the materially grounded empirical research process that he has here in mind. The scientist who treads new terrain always proceeds from an unclear initial situation. Like Bachelard, Fleck assumes that clarity in the empirical sciences is the effect of a work of clarification, in principle something belated. "If a scientific experiment were well defined, it would be altogether unnecessary to perform it. For the experimental arrangements to be well defined, the outcome must be known in advance; otherwise the procedure cannot be limited and purposeful. The more unknowns there are and the newer a field of research is, the less well defined are the experiments."[41] They only become more clear and consciously aimed, but for this very reason also more restricted, to the extent that "*they are carried along by a system of earlier experiments and decisions.*"[42] The direction that such a flow of experiments takes is not mapped out in advance. It depends on the "signal[s] of resistance" that emerge in the research process.[43] In connection with the history of the Wassermann reaction, Fleck speaks of an "edifice of knowledge . . . that nobody had really foreseen or intended" but that finally arose "in opposition to the anticipations and intentions of the individuals who had helped to build it."[44]

A style of thought is thus embodied in what I call an experimental system.[45] The constraints on thought that the construct of a system of this kind brings with it, the "dominant and directing disposition" that it produces,[46] exist in the form of material frameworks that are not all directly and consciously perceived by the scientist who has learned to move within them and work with them. They exist as a laboratory reality. Under the conditions of a developed system of this kind, the question naturally arises as to how major *changes* in style of thought can arise. Fleck does not have a very convincing response to this question, despite the fruitfulness of his comments on the *formation* of a thought style as the gradual internalization of an external signal of resistance. We shall return to this question again in connection with Thomas Kuhn. Fleck contents himself here with the rather sweeping conclusion that transformations in a style of thought, which he equates with "important discoveries" that bring about new fixations in relation to a previous style of thought, often coincide historically with times of general social unrest.[47]

Fleck defines the thought collective, complementarily with the style of thought, as a "community of persons mutually exchanging ideas or maintaining intellectual interaction," and who function here as *"carrier[s]' for the historical development of any field of thought, as well as for the given stock of knowledge and level of culture."*[48] Scientific discovery, for Fleck, is fundamentally a three-way relationship, the connection of the discovering individual and the object of discovery to an established state of knowledge. The closeness to Bachelard is again visible here, even if the latter, unlike Fleck, focused on stocks of knowledge principally in their embodied form of instruments. Knowledge acquisition as a social process, for Fleck, consists in the articulation of such stocks of knowledge. Patterns of knowledge are patterns of culture. Consequently Fleck sees the process of gaining scientific knowledge as the "most strongly socially conditioned human activity." It is accordingly for him the "paramount social creation."[49] It realizes itself by following its own dynamic, generally indeed to a certain extent even behind the back of the actors involved. Authorship, for Fleck, is thus located not in the individual scientist but rather in the collective, the "practice of cooperation and teamwork,"[50] as he says at one point.

Fleck's own historical case study consequently takes a corresponding form. His history of syphilis research, despite the eponym of the serum reaction that stands at the center of his analysis—the Wassermann reaction—is not concerned with big names. It gives a greater place to the working communities that strove with more or less success to bring a complex biological process into laboratory form. In this connection, Fleck made fruitful use of his own laboratory experience for the interpretation of the historical material, and in this way sought to tie together the two sources from which he intended to create his theory and history of knowledge—he himself speaks of a "history of knowing"—*Wissensgeschichte*.[51]

. . .

In ways that sometimes differ in degree, but not in principle, both Bachelard and Fleck worked for an epistemology that emphasized the social constitution of science as well as a specific historicity inherent in its pace. In Bachelard's case, the historical dimension was more to the fore, and in Fleck's the social. There was one point, however, where they fundamentally parted company. Ludwik Fleck tended to argue from the perspective of a body of knowledge that had already been granted legitimacy, at least for the time being, and that despite all the contingencies of the knowledge acquisition process aimed at a closure appropriate to its thought style and found its completion in an internally consistent "harmony of illusions" difficult to break up.[52] Gaston Bachelard argued more from the perspective of knowledge in the making for a mechanism of permanent revolution, by which in-built internal outbreaks maintained a constant state of provisionality and provided the cantons of knowledge with ever new configurations, thus keeping them in a state of permanent lack of closure. Fleck's perspective tended to be more pessimistic, while Bachelard's can be described as more optimistic in its assessment of scientific knowledge. While one inclined toward a critique of ideology, the other tended to a belief in science. Both worked, however, one from the perspective of research in physics and the other from that of research in biology, toward an alternative to logical positivism. They did indeed share with its representatives the intention to exclude questions seen as metaphysical from the philosophy of science. But while logical positivism aimed at explaining science in terms of the structure of its propositions, epistemologists such as Bachelard and Fleck displayed the attempt to grasp science as an epistemic and cultural process in which new knowledge may arise but at times may also be prevented.

3

Between the Wars—II

After the discussion of Bachelard and Fleck, the forms of historicization of epistemology that arose between the two world wars shall be followed in this chapter with a look at a number of other positions. The backgrounds and motives of their representatives were very different in nature, as likewise their philosophical provenance. What they have in common, however, is that they all contributed to sharpen the awareness of the need for a theory of knowledge that included an understanding of its development. All had to face the revolution in scientific knowledge that had taken place in physics in the first decades of the twentieth century, with repercussions also in biology and psychology, and they reacted in diverse forms and in various ways.

One of the positions that I would like to dwell on in some detail, though at first sight it might not seem so relevant to the concern with historicization, is Karl Popper's conception of the sciences and their dynamic, as developed in his debate with representatives of the Vienna Circle in the late 1920s and early 1930s. Popper, who initially trained as a joiner and, after studying philosophy and other subjects, worked as a secondary school teacher, began a critical dialogue in the late 1920s with Moritz Schlick and Rudolf Carnap, as well as with Herbert Feigl and Otto Neurath. Popper's first major book, published in 1935 under the title *The Logic of Scientific Discovery*, emerged from this debate. The philosophical tradition that it founded became known under the label of critical rationalism.

As a motto for his book, along with a line from Novalis, Popper chose a quotation from the historian Lord Acton, which reads: "There is nothing more necessary to the man of science than its history, and the logic of discovery . . . : the way error is detected, the use of hypothesis, of imagination, the mode of testing." At the very start of his book Popper defined the "logic of knowledge" as a logic of *research*: it deals with the methods of investigation and the dynamic of the empirical sciences they engender.[1] According to a very widespread misunderstanding, Popper contends, the method of scientific discovery is inductive in nature; and it is against this supposed logic of induction, whose difficulties he sees as impossible to overcome, that Popper develops his doctrine of the "*deductive* method of testing."[2] The basic issue for him is "*to formulate a suitable characterization of empirical science*,"[3] for he admits that he is "attracted by the adventure of science, and by discoveries which again and again confront us with new and unexpected questions, challenging us to try out new and hitherto undreamed-of answers."[4] From this perspective, it must at least seem remarkable that Popper then dismisses the core of discovery, the coming into being of new knowledge, completely from the horizon of his epistemological considerations, concentrating exclusively on the process of testing hypotheses, taking for granted their existence and not problematizing the way they emerge. This somewhat paradoxical constellation is worth examining more closely.

Popper's basic position is to be understood, according to his own admission, as an attempt to overcome positivism in the theory of knowledge. In its older form, he sees positivism as resorting to concepts of experience; in its more recent form, particularly that made popular by the Vienna Circle, to propositions of experience. Popper's argument is as follows: in order strictly to avoid a positivist epistemology, the problem of experience—with its necessarily subjective connotations—must be completely banned from epistemology. Experience in the sense of sensation and perception is a mode of the knowing subject. The same holds for the imaginative ability that is needed to think up a hypothesis. According to Popper, all these activities of the knowing subject belong in the realm of a psychology of knowledge, which must be kept sharply separate from the realm of a logic of knowledge. Popper defines the latter as being strictly a logic of scientific discovery or research (*Forschung*). It is about characterizing the *structure* of the research process that, in its formal character and transindividuality, opens an understanding of the progression of knowledge.

Popper naturally does not refrain from reclaiming the concept of experience for his own theory. He defines it for his purpose, however, in a radically new way. For Popper "experience" becomes an epistemological concept, a "particular method" that consists in a regulated procedure of "corroboration."[5] This method is the procedure of falsification, and it is here that Popper utters his famous sentence: "It must be possible for an empirical scientific system to be refuted by experience."[6] A system of this kind, which in the simplest case is a single universal proposition, is falsified if a concrete conclusion can be derived from it that does not stand up to experimental testing. Since an experiment is always a concrete and singular act, it can only contradict or provisionally confirm concrete and singular assertions. The logical procedure of the *modus tollens* makes it possible, however, to thus falsify a general proposition.

Even if their perspectives are very different, we are reminded here of Bachelard, who characterized scientific thought precisely as a thought being conscious of its own historical errors. Whether he and Popper were aware of each other's work at this time is unclear. They might very well have met in 1934 at the Eighth International Congress for Philosophy in Prague, which both attended. Bachelard, as we saw, considered truth scientifically as the historical rectification of a previous error, and experience as "a corrective for common and primary illusions."[7] But Bachelard's concept of rectification did not have the formal structure of Popperian falsification. His idea of rectification was neither that of inductive verification nor that of deductive falsification. It sought rather to grasp the directing force of experiment in the sense of a permanent reorientation.

Popper paid a high price for his solution, or rather for the way he dispensed with the problem of induction, in two respects. First of all, he had to renounce consideration of the most fascinating question that faces anyone who by his or her own confession is "attracted" to the adventure of scientific discovery, that is, the question of how *new* knowledge arises. This problem is relegated to the sphere of psychology and ascribed to what came to be termed the context of discovery, beyond the reach of a logic of knowledge. In Popper's epistemology, it is replaced by the question of how existing knowledge—however hypothetically it might have arisen—can be eliminated. Second, this involves a considerable impoverishment of the concept of experiment, which is reduced to an instance of testing hypotheses. An essential dimension of experiment is thus removed: its function in the exploration of new fields of knowledge; and its constructive role in the presentation of phenomena is bracketed out. Every next step on the path of knowledge is initiated by a bold hypothesis, which must be sufficiently explicit so that it can be tested. If for Bachelard empirical thought always only reaches clarity in retrospect, for Popper clarity is a precondition: the scientist must always already have an explicit conception of what might be rejected.

In later years, Popper placed his logic of scientific research in an evolutionary perspective, endowing it with the character of a mechanism of the growth of knowledge that drew on the theory of biological evolution. "The central problem of epistemology has always been and still is the problem of the growth of knowledge. *And the growth of knowledge can be studied best by studying the growth of scientific knowledge,*"[8] he wrote in 1958, in the preface to the first English edition of *The Logic of Scientific Discovery*. The theory of knowledge, therefore, is in the last analysis a theory of the growth of knowledge. In *Objective Knowledge: An Evolutionary Approach*, Popper explicitly describes the process he had depicted in his earlier work as a procedure of "discovery and elimination," with the formal structure of an evolutionary process based on variation and selection. He thus no longer speaks of a "logic" of scientific discovery, but quite consistently of a "biology."[9] In the context of Popper's theory of three worlds—the worlds of physical reality, of psychological reality, and of cultural reality—"objective knowledge" and its growth are viewed as a cultural product, and though it is a human creation, it also produces, as a collectively realized formation of a unique kind, a sui generis dynamic. It exists "to a large extent autonomously," and also generates its own problems, "especially those that are bound up with the methods of growth." In connection with his evolutionary conception and its strong emphasis on the rejection of mistakes, Popper even speaks on occasion of a "rational theory of emergence."[10]

The emphasis here is on a "rational theory," for what "emergence" means remains largely undeveloped. Popper however emphasizes the "objectivism" (*Objektivismus*) of his epistemology.[11] Individual scientists are far more stamped by the advancement of knowledge as an autonomous process than they are in a position to change it in creative ways. In the context of such a logic of evolutionary transcendence, a tendency already visible in *The Logic of Scientific Discovery* finds its continuation: that of understanding knowledge not as a subjective relationship but rather as an objective, constantly developing formation. Yet Popper's depiction of it, as we have seen, remains remarkably abstract. We shall return to the theme of an evolutionary logic of the process of knowledge acquisition in connection with the historical-epistemological writings of Thomas Kuhn and Stephen Toulmin. For a history of science aiming at a theory of knowledge acquisition, this has been a recurring option since the late nineteenth century.

. . .

While the philosophical outsider Popper was debating with the Vienna Circle in Austria, the philosophical outsider Bachelard had acquired his first professorship in Dijon and written *The New Scientific Spirit*, and the outsider Fleck in Lwów was seeking a publisher for *The Genesis and Development of a Scientific Fact* (Moritz Schlick replied in the negative), the emeritus mathematician and philosopher Edmund Husserl in Freiburg was working, against the background of the advancing wave of Nazism, on his late work *The Crisis of European Sciences and Transcendental Phenomenology*. An essay of his, "The Crisis of European Humanity and Philosophy," which Husserl presented to the Vienna Kulturbund in May 1935, formed part of the work on this book, as did a text published posthumously by Husserl's student Eugen Fink in 1939 under the title "Die Frage nach dem Ursprung der Geometrie als intentionalhistorisches Problem" (The Origin of Geometry) in the *Revue Internationale de Philosophie*.

In his last major work, and in particular in the two texts just mentioned, Husserl engaged intensively with problems of the history of science from his own perspective of a theory of knowledge. For him, this engagement was provoked by a crisis that he saw as essentially determined by the development of the natural sciences and the "positivist concept of science in our time" to which they had given rise. Husserl emphatically saw this as a "residual concept,"[12] which did justice neither to the constitution of the modern sciences nor to their history. For him, the essence of the crisis of the sciences of his day lay less in the most recent discoveries of physics, which with their antipositivist and antinaturalist philosophical resonances he could only welcome, to the degree that he was aware of them. The "greatest danger" for us "men of the present," in Husserl's diagnosis, was rather the hostility toward science that had become widespread especially in the young generation in the wake of the First World War, and that he saw as essentially determined by the positivist and naturalistic self-conception of the contemporary sciences with their technicized methods. European humanity had been deeply marked by the project of Western science, but was now being captured and threatened by this "residual concept" with its surrogate meanings. Western science had thus to be once again placed in its proper historical light: "We can gain self-understanding, and thus inner support, only by elucidating the unitary meaning which is inborn in this history from its origin."[13]

Husserl's argument can briefly be elaborated as follows. What we need is a nonpositivistic, nonnaturalistic understanding of the "historical event called 'natural science.'"[14] The perceived crisis was essentially the product of a self-understanding of the sciences that grew throughout the nineteenth century and cumulated in their self-presentation as "sciences of facts": "merely fact-minded sciences make merely fact-minded people."[15] On the one hand, the increasing technicality of methods in the exact sciences led so far away from the original comprehensibility of the scientific mode of obtaining knowledge, with its roots located in the horizon of our life-world, that "the primary and proper sense of science [could] be lost in favor of a substitutional sense. The technicized method proceeds by operating with thoughtless words and signs that have been emptied of their primary and proper meaning and mode of validity."[16] On the other hand, there was the increasing specialization of the individual sciences, which promoted a "no-longer-getting-involved with questions of meaning," although their respective areas of meaning were actually part of the meaning of the world, as Husserl put it.[17] However, neither the limitation of the realm of meaning nor the technicization of methods, according to Husserl, were given by the nature of the scientific procedure as such, let alone grounded in the natural properties of the world. Rather, they were the products of a process, the work of earlier generations of scientists, with a continuing historical becoming that should be kept alive—and which could only be assured and preserved as a cultural achievement whose understanding was beyond the technicized methods of the natural sciences:

It is also apparently quite forgotten that natural science (just like all science) is a label for intellectual achievements, in particular those of the collaborating scientists; as such, however, these belong like all intellectual achievements to the circumference that has to be explained by the humanities. Is it not then contrary to sense and circular to try to explain the historical event called "natural science" in terms of natural science, to explain it by bringing in natural science and its natural laws, which as an intellectual achievement themselves belong to the problem?[18]

The difficulty here is brought to a head: "The scientist is not aware of the fact that the constant foundation of his subjective mental work is the lived environment, which is constantly presupposed as the ground, the field of work, on which his questions and his method of thinking alone have any meaning. Where will the tremendous piece of method that leads from the perceptible environment to the idealizations of mathematics and their interpretation as objective being, be subjected to critique and clarification?"[19]

Husserl approached this "tremendous piece of method" in the first draft of his fragment on the "Origin of Geometry." If critique and clarification succeeded in the idealizations of geometry, he argued, then this should also be considered achievable in other sciences. The starting point of his inquiry was "the geometry, available to us through tradition."[20] For Husserl—as later also for Jacques Derrida, who translated and commented on this text of Husserl's—the development of forms of written fixation played a key role in this process of tradition. In this way, mathematics can assume, in the context of an "open chain of the generations of those who work for and with one another," the mode of being of "a total acquisition" as "the total premise for the acquisitions of the new level."[21] We have seen how this collective dimension of the production of knowledge, considered both synchronically and diachronically, took for Bachelard the form of an instrumental embodiment, at least as far as the physical sciences were concerned. For Husserl, on the other hand, writing becomes the bearer of a process of sedimentation, in which "scientific thinking attains new results on the basis of those already attained, that the new ones serve as the foundation for still others," for "all new acquisitions are in turn sedimented and become working materials."[22] Writing is thus responsible for the fact that the sciences are an "advancing formation of meaning,"[23] through time, for which the "*total* meaning" could not yet have been present "as a project and then as mobile fulfillment at the beginning."[24]

By way of the process of externalization through writing, something of the order of a creation of meaning in the historical process of the development of a science takes place—meaning is formed that can sediment and lead again to new problem states. Formation and sedimentation of meaning are interwoven. The possibility of such a displacement is the condition for the rise of genuinely new knowledge, not predictable at the start, and which binds every "subjective work of thought" into a generative process and one that spans generations, while on the other hand it also forms the starting point for that submersion of the "original and particular significance" which Husserl diagnosed, not only as a deficit of self-perception in the contemporary natural sciences, but also as an encompassing threat to European culture. This gives rise to the task of a "genuine history of philosophy," which Husserl mentions in the same breath with "a genuine history of the particular sciences," thus seeing them as one and the same task, namely: "the tracing of the historical meaning-structures given in the present, or their self-evidences, along the documented chain of historical back-references into the hidden dimension of the primal self-evidences which underlie them."[25]

A derivation of this kind precisely needs a historical epistemology. Without it, there is no prospect of a "depth-inquiry which goes beyond the usual factual history" and that is not reducible to historicist dogma.[26] Certainly, Husserl complains, epistemology has never been seen as a particularly historical task. But this is something with which we can precisely reproach the past, as "everywhere the problems, the clarifying investigations, the insights of principle are *historical*."[27] Epistemological grounding and historical explanation must coincide in the last analysis, they must ultimately be understood in a "regressive inquiry" (*Rückfrage*) as one and the same task, which at the same time is raised to the questions of our present, for this is, for Husserl, "what is historically primary in itself."[28]

An "internal history" of this kind, which goes beyond the mere "history of facts," rests for Husserl on a "universal historical *a priori*," and continues to have its ultimate foundation in the deep belief in a "universal teleology of reason."[29] As a specific challenge for any ambitious history of science since Husserl, the problem remains to explore the possibility of how to leave behind such a universal teleology and yet not to abandon altogether an order of historical reason, whatever form it might take.

. . .

In his 1938 essay "The Age of the World Picture," Martin Heidegger—who had been Husserl's student and had followed him in the chair at Freiburg in 1928—also engaged with the nature of modern science, more decidedly than he had already done ten years previously in *Being and Time*.

This text appeared—along with other essays from the period after Heidegger's notorious Freiburg rectorate—for the first time in 1950, in the collection *Holzwege* (*Off the Beaten Track*). It is astonishing how many parallels there are to be found between Heidegger's diagnoses and the conceptions of science from the 1930s that we have already discussed, especially with Bachelard's deliberations on the "new scientific spirit." The elements of Heidegger's diagnosis that are particularly relevant in the present context can be briefly summarized as follows.

Science, in the form that we have it today, in Heidegger's view belongs to the essential and striking phenomena of the modern age. It made its appearance together with another phenomenon of equal importance: machine-based technology. The two things are related in the sense that machine technology as a particular form of the practice of human labor challenged and brought the natural sciences into being, eventually demanding the development of their mathematicized form. Thus it was not natural science that presided over the birth of modern technology, by its calculation and quantification. On the contrary, technology is the basis from which natural science, in its characteristic modern form, derives its driving momentum. This particular constellation, according to Heidegger, is qualitatively new in relation both to the Greek episteme and to the medieval doctrinaire form of science, and is characteristic of the modern age as a whole. There is no continuous development from one form of science to the other, but rather an epochal break between them. Technology, for Heidegger, also founds the essence of modern metaphysics. The question as to what comprises the essence of contemporary *science* was answered by Heidegger as follows: "The essence of what we today call science is research. In what does the essence of research consist? In the fact that knowing [*das Erkennen*] establishes itself as a procedure within some realm of what is, in nature or in history. Procedure does not mean here merely method or methodology. For every procedure already requires an open sphere in which it moves. And it is precisely the opening up of such a sphere that is the fundamental event in research."[30] In scientific discovery, knowing becomes a "procedure" in a space with an open horizon. Research discloses certain regions of objects according to a "project" (*Entwurf*), which at the same time also determines the kind of "rigor" to which research commits itself. "This binding adherence, the rigor of research, has its own character at any given time in keeping with the project."[31]

Like Bachelard, Heidegger sees the modern scientific spirit as being realized in the *Entwurf*—Bachelard's term, we remember, was "project"—and essentially developing in "spheres"—what Bachelard calls "cantons." These open spheres each display their own character of rationality, which must be understood on the basis of the specific procedural conditions in each of them. In this context, "rigor" or "exactness" is not an absolute degree of precision, but is rather dependent on the particularities of the objects that can be disclosed in the sphere in question, in which the procedure makes these objects accessible, and makes them intelligible "by calculation" through "the constant comparing of everything with everything."[32] Heidegger's objects conformable to a project can be compared to Bachelard's "techno-phenomena." Still another parallel can be noted here. Like Popper, Heidegger sees the nature of contemporary science in research. But where Popper seeks to grasp it in the form of a "logic," Heidegger sees in it a secular process closely bound up with material technology.

In the "becoming objective" of a region of knowledge the possibility arises of ever new knowledge coming into view. Here the "procedure," assured by rigor in the context of a project, can remain open to "the changeableness in whatever encounters it." New knowledge is realized by experiment, a form of obtaining knowledge that according to Heidegger "only becomes possible where and only where the knowledge of nature has been transformed into research."[33]

There is a further striking parallel between Heidegger and Bachelard. Like his French contemporary, Heidegger also does not see the specialization of the modern natural sciences as the deficient mode of a scientific spirit ideally characterized by an encompassing understanding, but rather as a structural particularity of modern knowledge based in the nature of the procedure. "Every science is, as research, grounded upon the projection of a circumscribed object-sphere and is therefore necessarily a science of individualized character." And he continues: "This particularizing (specialization) is, however, by no means simply an irksome concomitant of the increasing unsurveyability of the results of research. It is not a necessary evil, but is rather an essential necessity of science as research."[34] Heidegger identifies the advantage of this system that it, though "regulated," has a high flexibility by virtue of its ramification,[35] and makes it possible also to respond to new challenges in a flexible way.

This network that tends to separate into special knowledges is held together by something that Heidegger describes as the basic procedure of the "pursuit." *Pursuit* here does not mean either assiduousness or some kind of arrangement of ongoing production. By pursuit Heidegger understands the recursive cohesion that modern natural science obtains and maintains by drawing on technological practice and returning its results back into that practice. One might be hearing Bachelard once again when Heidegger states: "The methodology through which individual objectspheres are conquered does not simply amass results. Rather, with the help of its results it adapts [*richtet sich . . . ein*] itself for a new procedure. Within the complex of machinery that is necessary to physics in order to carry out the smashing of the atom lies hidden the whole of physics up to now."[36] Instruments, Bachelard says, are embodied theorems, whose deployment can in turn lead to techno-phenomena that shift the theoretical axes of the experiment under which they took shape. The recursive loop that is thus given, in which the results of the procedure become the precondition for its productive continuation, lend modern science its particular systemic character, which in the last analysis can be understood only in its temporal dimension. It has a historic path that can best be grasped with the concept of recursive differentiation: "More and more the methodology adapts itself to the possibilities of procedure opened up through itself. This having-to-adapt-itself to its own results as the ways and means of an advancing methodology is the essence of research's character as ongoing activity."[37]

This process, for Heidegger, also encloses the core of contemporary metaphysics. Its concept of being, in particular, is the concept of the object in modern science: "Only that which becomes object in this way *is*—is considered to be in being. We first arrive at science as research when the Being of whatever is, is sought in such objectiveness."[38] "What is, in its entirety, is now taken in such a way that it first is in being and only is in being to the extent that it is set up by man, who represents and sets forth."[39] "From now on," as we read in Bachelard, "hypothesis is synthesis."[40] Heidegger's later critique of technology is certainly latent in "The Age of the World Picture," but it is in no way anticipated as a critique. It is clear, however, that, unlike Popper and also as distinct from his teacher Husserl, both of whom saw the essential achievements of science in ideal forms (theories, concepts), Heidegger understood the particularity of modern science precisely as a material form of the mobilization of rationality, as an "applied rationalism,"[41] to use once again a phrase of Bachelard's, and as having an essentially collective constitution. Contemporary scientific "representation," for Heidegger, consists in a quite literal sense of a "representing and setting before [*Vor-sich-Hinstellen*]," which, as he himself puts it, results in a "structured image [*Gebild*],"[42] that is the creature of man's producing, and thus in a deep-reaching transformation of our life-world.

. . .

At this point, finally, we must take a look at Ernst Cassirer. Coming from the Marburg school of science-oriented neo-Kantianism, Cassirer taught from 1919 at the newly founded University of Hamburg. In 1933 he emigrated to England, then to Sweden and eventually to the United States, where he died in 1945. Alongside Husserl, Cassirer was one of the few twentieth-century philosophers in Germany who consistently worked to prevent the natural and cultural sciences of their time from completely falling apart, conceiving them as a differentiated, historically articulated framework of cultural knowledge—expressive, depictive and symbolic—in which each side ultimately referred to and depended on the other.

In the context of a "philosophy of symbolic forms," which was Cassirer's ultimate goal, natural science can only be seen as a "special case of objectivization in general,"[43] and more closely as a case that appeared at a late date in human history. According to Cassirer, all cultural phenomena of objectification pursue grosso modo the path from the sphere of expression to the sphere of representation, and thence to that of symbolic meaning. In the conceptual form of the natural sciences, Cassirer believed he could particularly distinguish a "mimetic" from an "analogical" phase, which ultimately gave way to the "properly symbolic form of concept formation." To the extent that the development of science can be inferred from its reflex in various contemporary philosophical systems, the names of Aristotle, Descartes, and Leibniz can stand for these phases.[44]

For the nineteenth and twentieth centuries, according to Cassirer, this form of reflection, so convenient for philosophers, is no longer possible, since "there is no longer any great representative philosophical system in which we can discern the status of scientific theory and methodology."[45] Cassirer took up this challenge and pursued it intensively, discussing in detail recent developments in the sciences themselves, not just in physics but also in biology. In the last volume, published posthumously and initially in English, of his major work *The Problem of Knowledge: Philosophy, Science, and History Since Hegel*, the first volume of which had been published already in 1906, he goes still further and ascribes the fact that today there are as many individual theories of knowledge as fields of research to "the real, inner, moving forces" that lie "often deeply hidden, within the sciences"—again not so unlike Bachelard. Their understanding consequently requires a "patient steeping of oneself in the work of the separate sciences."[46] Certainly, "the era of the great constructive program, in which philosophy might hope to systematize and organize all knowledge, is past and gone." But this did not mean for Cassirer any renunciation of reflection. It was rather that, in place of earlier philosophical systems with their integral metaphysics, we need an intensive debate with the regional formation of concepts, to which the sciences find themselves driven in the particular realms of their development. "The demand for synthesis and synopsis, for survey and comprehensive view," he went on, "continues as before, and only by this sort of systematic review can a true historical understanding of the individual developments of knowledge be obtained."[47] To do justice to this "genuine historical understanding" requires a cultural history of knowledge that follows its modern ramifications and takes them seriously in their various manifestations.

Cassirer became increasingly aware that the recoinage of modern knowledge in its particularities could not be understood simply on the basis of the results it gave rise to. "We must not content ourselves with considering its product; we must investigate from within the mode and direction of its production."[48] For evidently what was involved in the scientific appropriation of the world was "not a matter of disclosing the ultimate, absolute elements of reality, in the contemplation of which thought may rest as it were, but of a never-ending process through which the relatively necessary takes the place of the relatively accidental and the relatively invariable that of the relatively variable."[49] Objectivity no longer presented itself as a given, but rather as a "task" of objectification, in which theoretical knowledge remained bound to the forms of its cultural-historical, material objectifications. Conversely, however, it had also to be admitted that "all theoretical concepts bear within themselves the character of 'instruments.' In the final analysis they are nothing other than tools, which we have fashioned for the solution of specific tasks and which must be continually refashioned."⁵⁰ Cassirer's "task" of the modern production of knowledge can be compared with both Bachelard's "project" and Heidegger's "*Entwurf.*" As Cassirer stated programmatically in his *Logic of the Humanities,* "every energy of the mind shares in it" in its own peculiar fashion.⁵¹

In these late studies, published in Gothenburg in 1942, Cassirer defined a cultural object as follows:

> Like every other object, an object of culture has its place in space and time. It has its here-and-now. It comes to be and passes away. Insofar as we describe this here-and-now, this coming-to-be and passing-away, we have no need to go beyond the sphere of physical determinants. But, on the other hand, in this description even the physical itself is seen in a new *function.* It not only "is" and "becomes"; for in this being and becoming something else "emerges." What emerges is a "meaning," which is not absorbed by what is merely physical, but is "embodied in and through it"; it is the factor common to all that content which we designate as "culture."⁵²

A thorough cultural history of the sciences will thus have to take issue with embodied meanings of this kind, that is, with specifically scientific meanings. In his five studies, however, Cassirer did not analyze the sciences in any further detail, although he did generally indicate that physical, historical, and psychological categories had to be brought into a synthesis, if the description of a cultural object was to be successful. The historical, in this connection, is not simply what has been, but "possesses and retains a present peculiar to itself,"⁵³ a present that poses, to take up Husserl's formulation again, the "historically primary in itself."⁵⁴

In the eyes of Cassirer, the scientific present lent the cultural sciences unexpected protective help from the side of the natural sciences. In terms of method, the cultural sciences had been in an uncomfortable position as long as the "mechanical world view" of the natural sciences remained unchallenged. "But it was just here that that remarkable development occurred which led to an inner crisis and finally to a 'revolution in our mode of thinking' within the field of the science of nature."⁵⁵ Cassirer saw the core of this revolution—at the very heart of natural-scientific thought—in the rehabilitation of particular concepts of wholeness and structure. This in no way meant for him a misunderstood license to efface the boundaries between the natural sciences and the humanities; rather, the latter could now "abandon [themselves] to consideration of [their] own forms and [their] own structures with greater freedom and less restraint than heretofore as a consequence of the fact that the other fields of knowledge have become more mindful of their own very real problems of form. The logic of research is able now to assign each of these problems to its rightful place."[56] Although the logic of research could now ascribe all these problems their place, for Cassirer, a difference persisted in the "logic of research" between the natural sciences and the humanities. This was the role played by history in the understanding of their respective cultural objects. But the natural sciences, considered in their historical development, could now be seen as one of the privileged objects of modern culture, and hence of the humanities. The logic of their exploration was thus placed on the agenda, even if Cassirer himself had done not much more than indicate this as a project in its own right. In Chapter 5, we shall look in more detail at the difference in the constitution of the objects in the natural sciences, and in the history of science as concerned with the natural sciences, in particular on the occasion of a review of the writings of Georges Canguilhem.

4

After 1945

In the first few decades after World War Two, a further decisive change took place in the relationship between philosophical and historical reflection on the sciences and their development. I would describe it as the transition from attempts at historicizing epistemology to attempts at epistemologizing history of science. The first half of the twentieth century was dominated by considerations that had their origin in philosophical questions arising from contemporary developments in the sciences but remaining bound to the traditional inventory of the problem of knowledge; what now emerged was a historical assessment of the course of the sciences in the form of a new reflection on method bound up with that assessment. This turn would have decisive significance for the relationship between both the history and the philosophy of science in the second half of the twentieth century. So far, however, it has been little studied and scarcely understood—especially the sweeping interest of its peers in the early modern age in the midst of a present of political upheavals that ended in World War Two followed by the Cold War; a present, moreover, in which nuclear physics came to be seen and experienced no longer as simply an adventure of knowledge but as the basis for a new form of mass destruction. Only today is the history of science becoming aware of its own involvement in a global history that does not leave its preferences and procedures undisturbed.

An important figure in this transition was the French historian of science Alexandre Koyré. Born in Rostov and having grown up in Russia,

Koyré studied mathematics and philosophy at Göttingen, where David Hilbert and Edmund Husserl were among his teachers, then in Paris, where in the early 1930s, under the influence of Émile Meyerson, he turned toward history of science. His *Galileo Studies*, published in Paris in 1939, brought a new understanding of Galileo's epoch, the late sixteenth and early seventeenth centuries, which Koyré described as a "scientific revolution." Koyré reached the United States via Cairo in 1941, where after the war he influenced a whole generation of historians of science—including Thomas Kuhn and Bernard Cohen—who aimed at transforming the history of science into a respected academic discipline. In America, Koyré published in 1957 his major work *From the Closed World to the Infinite Universe*.

In a wide-ranging synthesis of history of philosophy and history of science—essentially from the perspective of a history of ideas—Koyré sought to reveal what he called the "structural schemas of the old and new world view." He identified them neither with the change from a science of contemplation to one of intervention, nor—like Heidegger—with the introduction of machine technology, nor again with the replacement of transcendence by immanence. The underlying level for Koyré was rather a radical change in worldview: the destruction of the closed cosmos handed down from antiquity and the concomitant geometrization, that is, the opening up of space in the direction of a universe of infinite extension.[1]

Koyré's history-of-ideas approach, oriented to world pictures and basic assumptions, and to a certain degree even antiexperimental, became characteristic of the history of science in the 1940s and 1950s, but was to be subjected to a persistent critique that we will later discuss. Koyré, as he readily acknowledged, saw "scientific work as above all theoretical work."[2] It was also perhaps no accident that the shift in the center of gravity from the philosophy of science to the history of science, of which Koyré was an exemplary representative, was accompanied by an emphasis on the theoretical in science, since it revalorized the turn toward history itself. A theoretically demanding historicization thus contested the field of traditional philosophizing about science. It was important for the contemporary reception of Koyré's writings that he maintained the existence of something like a revolutionary *break*—even if it extended over more than two centuries—which eventually separated two cosmological conceptions from one another, in which not only scientific factors but also ideas "of a philosophical, magical and religious nature"—factors "beyond logic,"[3] therefore—were in each case coherently interrelated. For Koyré, however, the heart of the matter was the transition from a "physics of experience and sensory perception" to a study of natural phenomena in which "mathematics had precedence."[4] What appeared clear and self-evident to Galileo and Descartes would have been seen as quite obviously false in antiquity or the Middle Ages, if not indeed as absurd. Koyré concluded: "It is only possible to explain this if we admit or recognize that all these 'clear' and 'simple' concepts that provide the foundation of modern natural science, are in no way so in and for themselves, but only in connection with broader ideas and principles, without which they are by no means 'clear.'"[5]

In the same sense, Koyré sought to break with the custom in the history of science, apparently so self-evident, of viewing the scientific achievements of the past in the uninterrupted light of the present. From such a perspective "we do not recognize Galileo's rashness in deciding to treat mechanics as a branch of mathematics, that is, to replace the real world of everyday experience with a merely proposed reality of geometry, and explain the real in terms of the impossible."[6] Koyré practiced and taught a view of the history of science that was prepared to engage with the conditions, to be investigated historically, that the early modern age had left behind with the rise of the modern natural sciences. The fact that this view took shape at the very point in time when the era of classical physics—an expression, incidentally, which Koyré rejected for modern physics—had reached a key turning point, makes the relationship of historians of science to the objects of their investigation appear more complicated than the historically limited self-understanding of those historians, for whom it was the very badge of honor of their profession that they viewed their objects exclusively in the light of the time that they were investigating. Even Herbert Butterfield—who in his book *The Whig Interpretation of History* championed Koyré's depiction of the scientific revolution, seeing the historian as attempting "to see life with the eyes of another century than our own"—did not simply reject study of the "past in relation to the present" tout court, insisting rather that the unavoidable implications of this relationship should be "carefully considered."[7]

· · ·

Thomas Kuhn was one of the historians of science of the postwar generation who consistently took up and pursued further the idea of breaks in scientific thought, which Koyré had not been the first to spell out but had depicted in broad lines and to great effect. Kuhn had studied physics at Harvard before he turned to the history of his discipline. He began the introduction to *The Structure of Scientific Revolutions*, published in 1962 and certainly his most influential book, with an observation and a challenge. As Kuhn put it, the historiography of science could change our present standard image of the sciences, if it was not limited to repeating anecdotes and tabling chronologies. Up to now, this image had generally centered on the depiction of established scientific achievements as they were found in the textbooks. For Kuhn, this image was decidedly erroneous. Accordingly, he defined the aim of his essay as "a sketch of the quite different concept of science that can emerge from the historical record of the research activity itself."[8]

Kuhn ascribed primary responsibility for the traditional image of science to its presentation in academic textbooks: practicing scientists, including most of those who wrote textbooks, did not generally take the time to consider the history of their profession—and by their academic socialization and experience of writing they were generally not in a position to do so. Those who were methodical historians, however, for the most part created their image of the sciences from precisely this kind of literature, thus confirming the same prejudice.

Kuhn set out to break through this circle, comparable in this way also to Fleck, with whom we saw a similar argument about the "textbook form" of our customary image of science. The methodological lever that had to be used to prize open the circle was a break with the idea that science develops in the form of a more or less continuous "accumulation" of results. Wherever earlier historians of science had engaged in serious probing and gone into historical detail, this "piecemeal process" of growth had always proved inadequate. Kuhn saw signs of this inadequacy first of all in the difficulties of localizing supposed discoveries unambiguously in time and in relation to the scientists involved—whether this was a case of a new element such as oxygen or a principle such as the conservation of energy. He saw the second difficulty generally as consisting in having to decide where the "error" of the past ended and the "truth" of the present began. Perhaps, he wrote, "science does not develop by the accumulation of individual discoveries and inventions."[9] The more carefully we look, the more it turns out that the old ideas were no less plausible in their own time than their successors in ours.

Science is thus precisely not, we could say with Kuhn and against Popper, a continuous succession of falsifications—even if they concern discrete hypotheses that do not necessarily follow from one another. Even Bachelard's epistemological breaks would still have, for Kuhn, too much the spice of an overarching perspective of progress to support his own postulate of discontinuity. According to Kuhn, we have so far posed the wrong questions. A revolution was needed in the very way that the basic methodological assumptions of the history of science were conceived. "The result of all these doubts and difficulties," wrote Kuhn, "is a historiographic revolution in the study of science, though one that is still in its early stages."[10] Among the historians of science whose studies prepared this kind of revolution in historiography, and the possibility of "a new image of science," the only one Kuhn explicitly mentions is Alexandre Koyré. *The Structure of Scientific Revolutions* sketches the outlines of this new historiography. In the same year of 1962, the art historian George Kubler published a similarly radical book on recurring forms and discontinuities in art and architecture with the title *The Shape of Time*.[11] Conceived as notes toward a "history of things," Kubler's essay, however, found less resonance than Kuhn's, whose book was received as a successful attempt at historizing scientific modes of thinking.

Kuhn described his own procedure as interpretative and normative, that is, in no way purely descriptive, in the sense often demanded as the methodological ideal of historiography. He saw his conclusions as touching at the heart of epistemology. "Traditionally," he remarked, they "belonged to logic or epistemology"; in particular, they broke down the barrier between the context of justification and the context of discovery that had been established and cemented in analytic philosophy of science in the wake of Hans Reichenbach.[12] Popper's logic of scientific discovery had still accepted this barrier; while Popper had distanced himself from analytic philosophy of science with its antimetaphysical and linguistic orientation, he had still used the distinction as a demarcation for his own perspective. But Kuhn was also suspicious of another barrier. Like Fleck before him, he did not shy from enriching his historical logic of discovery with elements from sociology and social psychology. Kuhn thus presented a hybrid mix of epistemological, psychological, and sociological elements, about which he himself raised the rhetorical question: "Can anything more than profound confusion be indicated by this admixture of diverse fields and concerns?"[13]

We need to examine more closely how Kuhn redeemed this hybrid claim and manipulated the "admixture of diverse fields and concerns." Kuhn's starting point was that modern science is a collective enterprise pursued by a community of scientists, or a number of such communities, who at a given point in time take for granted a particular scientific achievement as the foundation for their actual work. Such scientific acquisitions have to show two characteristics. First, they must be attractive enough to win over a sufficient number of supporters for a substantial period of time. Second, they must be open enough for the community to relate to them a spectrum of unsolved problems, in the belief that these can be solved in the context of the acknowledged scientific achievement. An acquisition that shows these two characteristics is described by Kuhn as a "paradigm." Work carried out in the context of such a paradigm he refers to as "normal science."[14] Normal science can be seen as a kind of puzzle solving, as fitting together a jigsaw: the unit by which normal science is measured is the solved problem, the result, or the finding, and the progress of science is measured in this connection by the number of solved problems and established findings.

Now it can happen that in the context of such a paradigm, a certain problem stubbornly resists solution. As long as such an occurrence remains isolated, it can be considered as an anomaly within the prevailing paradigm and be characterized as an exception by means of additional ad hoc hypotheses. But a point may eventually be reached at which anomalies and exceptions accumulate to such an extent that this kind of quarantine is no longer sufficient, and confidence in the paradigmatic knowledge shrinks. In such a critical situation, when the old paradigm has lost its organizing power and the new order of knowledge has not yet established itself, a revolution may take place—similar to the sense of the classic political concept. A new guiding idea, if it is to hand at the critical moment, has the opportunity of replacing the old. A "paradigm shift" occurs. Two conditions however need to be fulfilled for such a shift to take place: "First, the new candidate must seem to resolve some outstanding and generally recognized problem that can be met in no other way. Second, the new paradigm must promise to preserve a relatively large part of the concrete problem-solving ability that has accrued to science through its predecessors. Novelty for its own sake is not a desideratum in the science as it is in so many other creative fields."[15]

In establishing these necessary conditions, the scientific community plays a decisive role. Kuhn believed that the structure of such communities had been far too little studied. When it is a question of a *scientific* revolution, the decision about a paradigm shift is a matter for this collective, and not for other authorities—whether religious, political, or whatever. "Just how special that community must be if science is to survive and grow may be indicated by the very tenuousness of humanity's hold on the scientific enterprise. . . . The bulk of scientific knowledge is a product of Europe in the last four centuries. No other place and time has supported the very special communities from which scientific productivity comes."[16] The sociological and sociopsychological elements that play a particular role in the context of a scientific revolution are thus for Kuhn closely limited, a fact that was often overlooked by his followers. What Kuhn thus has in mind are not external influences in the sense of an externalist historiography, but rather problems that arise in the context of a paradigm itself, which start the ball rolling and can lead to the replacement of the former paradigm.

Fleck had already reflected intensively on the formation of such scientific communities, which he described as thought collectives, and especially on the social and psychological aspects that play a role in their constitution and still more so in their perpetuation. But while Kuhn emphasized above all the question of how belief was shattered within a scientific community, leading to the conditions for a change in paradigm, Fleck had focused his attention on how thought collectives of this kind come to achieve perfection of a style of thought, in what he eventually described as a "harmony of illusions."

A paradigm shift is like a gestalt switch. What previously made sense no longer does so, and vice versa. What we are faced with here, accordingly, in the context of the displacement of one paradigm by another, is a threshold phenomenon, something like historical "incommensurability." It is questionable, therefore, and basically unreliable, to speak of an advance from one paradigm to the next, and better to abandon the idea that each shift of paradigm leads us "closer to the truth."[17] Not only are there historical discontinuities; according to Kuhn, the entire field of the sciences consists in its synchronic extension of so many scientific communities, which are fixed on their respective paradigms and conduct their work under mutually incommensurable assumptions. We thus see that Kuhn's work, too, is guided by the two sets of problems that we have identified as driving moments of twentieth-century reflection in philosophy and history of science: the historical, epochal relativization of the classical timeless ideal of science, and the problem of the unity of the sciences—that is, their irreducible plurality.

Kuhn gave the concluding chapter of his book the title "Progress Through Revolutions," which seems paradoxical after what we have said about his work. But he did indeed see something like an advance in the course of the sciences, even in the sense of a process of development: "Successive stages in that developmental process are marked by an increase in articulation and specialization." The whole process, Kuhn says at this point, "may have occurred, as we now suppose biological evolution did,"[18] though not without immediately warning against pursuing this analogy too far or taking it literally. In no way does he postulate that specific mechanisms effective in one domain should be effective also in others, contrary to what we saw in the case of the late Popper. What Kuhn is specifically concerned with is, rather, abolishing the idea that the development of knowledge involves a teleological moment: "The developmental process described in this essay has been a process of evolution *from* primitive beginnings—a process whose successive stages are characterized by an increasingly detailed and refined understanding of nature. But nothing that has been or will be said makes it a process of evolution *toward* anything." The question is to replace an "evolution-toward-what-we-wish-to-know" by an "evolution-from-what-we-do-know."[19] In a later reflection on his "historical philosophy of science," as he called it, Kuhn pertinently described the development of the sciences as a process "driven from behind,"[20] which therefore is and remains open ahead, in a way that escapes prediction.

• • •

Stephen Toulmin also grappled with an evolutionary model of scientific development in his writings on the philosophy of science. Toulmin studied mathematics, physics, and philosophy at Cambridge, where he conducted his doctoral studies under the influence of Ludwig Wittgenstein. In *Foresight and Understanding*, published in 1961 and subtitled *An Enquiry into the Aims of Science*, Toulmin—like Kuhn—explicitly referred to the theory of biological evolution. It became central to Toulmin's reflections on the course of the sciences that scientific rationality cannot be understood after the model of a formal argument, or by analogy with a logical schematism. He sought for models that could help him find a substitute: "Science is not an intellectual computing-machine: it is a slice of life."[21] Attempts to judge scientific theories by their prognostic power, and to derive a criterion of scientific progress from this, appeared to Toulmin—like all other attempted definitions resting on one-sided criteria—as insufficient, ultimately touching only the surface of things. The "rational patterns of connection" of an age, and long-term changes in the "ideals of natural order" operative in scientific research, form a far more complex connection, which in the last resort can only be made accessible to understanding by extensive historical analysis of both its structure and its dynamic.[22]

"For the logician," wrote Toulmin, "these explanatory ideals pose a particular problem. On the one hand, they change and develop, as time goes on, in the light of discovery and experience: so they must be classed as 'empirical,' in a broad enough sense of the term. On the other hand, one cannot confront them directly with the results of observation and experiment. They have to prove their worth over a longer term, in a way which still needs analyzing."[23] But this paradox of a certain historical a priori, as we could call it, lies at the core of any evolutionary process in Toulmin's understanding. He explains the paradox by a comparison: "There is only one way of seeing one's own spectacles clearly: that is, to take them off. It is impossible to focus both on them and through them at the same time. A similar difficulty attaches to the fundamental concepts of science."[24]

To stay with this image, the decisive question is how to take the spectacles off. What has to happen for them to be noticed at all? On this question, Toulmin parts company on the one hand with Popper's evolutionary logic, which closely coupled the refutation of hypotheses with observation and experiment, basing the procedure on the assumption of a conscious bearer of well-defined spectacles that may at any time be replaced by others. On the other hand, he takes his distance from Kuhn, whose revolutionary model Toulmin counterposes with an unambiguously gradualist one resting on steady small changes. Toulmin sees science as an ensemble of ideas and techniques, whose components and goals constantly change in a mobile intellectual and social environment. It is thus not enough for a historian to be a chronicler, or for a philosopher to retreat to a formal theory of science. Above all, it is not a question of any "single, simple test of merit, and it is not for the philosopher to impose one on science; nor can a historian justly criticize earlier scientists for not jumping straight to the views of 1960."[25]

A historical philosophy of the sciences should thus not be normative, and in particular should not universalize the standards of its own time. The task that confronts historians and philosophers of science, if they are to be "effective and realistic," has its parallel for Toulmin in "Darwinian biology."[26] Yet this parallel should not be reduced to an abstract principle of variation and selection; rather it should be considered in the perspective of a complex historical ecology. This particularly means bearing in mind two things. First, every environment is an ensemble of several factors, which may make quite different demands on knowledge. The principle of selection is thus itself a composite function. Second, procedures that prove advantageous in one particular constellation may be disadvantageous in another, perhaps even harmful and inhibiting, both synchronically, with respect to different fields of science that simultaneously coexist, and diachronically, with respect to the development of certain of these fields. In order to understand particular historical developmental processes, there is no alternative but to pursue detailed investigations. Empirical inquiry is inescapable, even though historical epistemology is well advised to pursue such research with an ecologically motivated idea of evolution in mind. Yet it remains impossible to foresee how the development of science will continue in the future.

Even if in his later book *Human Understanding* Toulmin spoke of the change of ideas in particular scientific formations as the "result of a double process of variation in ideas and intellectual selection," he leaves no room for doubt that both variation and selection involve complexly structured components, both "rational" and "causal."[27] These, in turn, underlie a constant process of historical revaluation. For Toulmin—as for Bachelard—

it is especially at the "front lines of reason" that the problems tackled in a field of research are "intrinsically 'cloudy.'"[28] Decisions are generally justified only ex post facto. Unlike Kuhn, Toulmin does not see a field of knowledge as maintained by a paradigm. Rather, "the intellectual strategies employed by the scientists working in any field, at one time or another and in one research center or another, themselves form an overlapping population, with its own internal variety."[29] The coherence of a given field as well as its margins are thus constantly in question.

. . .

From Fleck to Kuhn, from Popper to Toulmin, we find two common features despite all their differences: an epistemology that takes history seriously and does not just follow self-defined normative principles is confronted, first of all, with a pluralism of methods in the spectrum of the sciences—a pluralism which epistemology must find a form to cope with, instead of reducing it to a fictional unity. Second, it has to take into account that ideals of scientific method can change in the course of history, and that what is or is not taken seriously as scientific knowledge can actually vary. Paul Feyerabend was unique among theorists of science in the postwar period in his radicalization of these tendencies suspected as relativistic from an analytic perspective. His early studies had been in theater and astronomy, before he turned to philosophy. Feyerabend did his doctorate in the early 1950s in Vienna with the sole representative of the Vienna Circle surviving there, Victor Kraft, and was then for a while an assistant to Popper in London, before he eventually took a definitive turn away from Popper's critical rationalism. In his book *Against Method*, which followed a 1970 essay of the same title in *Minnesota Studies in the Philosophy of Science*, Feyerabend pleaded for an epistemological anarchism, though one that at its core made use of arguments from the history of science.

As a refrain to the title of his book *Against Method*, Feyerabend starts off by comparing Lenin's reflections on the history of political revolutions and the ideas of the historian Herbert Butterfield on the needlessness, or unreliability, of a victors' history, in order to explain his own methodological premises. In view of the complexity and unpredictability of the process of historical development, any methodology that appeals to principles is bound to end up appearing naïve. In any case, Feyerabend argues, it can tell us nothing about how "successful *participation*" in such processes is possible. Science is no different from politics in this respect: only "a ruthless opportunist who is not tied to any particular philosophy and who adopts whatever procedure seems to fit the occasion" can be successful in the process of scientific research.[30] The deepest reason for this is the radical contingency of history. In the last resort, therefore, it is always only a question of the "means for moving from one historical stage to another."[31] The history of science is a complex reality, in which successful intervention is possible only by a similarly complex activity, not on the basis of rules laid down in advance. We thus have a process in which being successful means nothing more than reaching the "next stage."

Feyerabend's argument is not as naïve and rhetorical as it may seem at first sight. He saw the epistemological anarchism that he promoted as a "remedy for strong minds," in no way a plea for general arbitrariness. "We do indeed want to create connections between well-articulated totalities, and not just a few waves in a sea of barren movement."[32] Not so dissimilar to Toulmin on this point, Feyerabend argued for recognition of the rich variety of science as a human reality—as against what Toulmin had called a "computing-machine"—a reality whose simplification by any theory of science he battled against with all the resources of argument at his disposal. His conviction was that a rationalistic simplification would eventually damage science by impeding the creativity of its practitioners. The only principle that he acknowledged—as a principle "that does not inhibit progress"—was a motto that subsequently became famous: "Anything goes."[33] Feyerabend directed his polemic against the "simplified models" of theories of science, which pretended to speak in the name of "that complicated, fine-spun, semi-intuitive and always contradictory model" that in his view was what made up the real existence of science. He wanted to defend this latter model, in all its complexity, against the "logicians."[34]

This is not the place to consider Feyerabend's scientific anarchism as a whole, but simply his particular contribution to the historicization of epistemology. Feyerabend emphatically maintained that scientific developments that lead to really new knowledge do not start, as Popper assumed, with precise ideas. It is the same with the objects of science as with freedom: "We must expect, for example, that the *idea* of liberty could be made clear only by means of the very same actions, which were supposed to *create* liberty. Creation of a *thing*, and creation plus full understanding of a *correct idea* of the thing, *are very often parts of one and the same indivisible process* and cannot be separated without bringing the process to a stop."[35] Feyerabend argues for an understanding of the empirical sciences and their modes of knowledge production that we first encountered not just with Toulmin but already with Fleck, and especially with Bachelard—in his non-Cartesian epistemology. In *The Formation of the Scientific Mind*, Bachelard contended: "Reality is never 'what we might believe it to be': it is always what we ought to have thought. Empirical thought is clear *in retrospect*, when the apparatus of reason has been developed."[36] The new, in this conception, does not come into the world as a clear and distinct idea, but rather brings itself into relief ex post facto, as it were. It results from a recursive process. Clarity is a historical product of the purifying work of the scientific mind, and not of its presumed Cartesian constitution. The empirical sciences *obtain* the concepts that are relevant for them in the experimental creation of their objects, they do not just apply them to a piece of reality.

The general rule, so Feyerabend puts it, is that "unreasonable, nonsensical, unmethodical foreplay . . . turns out to be an unavoidable precondition of clarity and of empirical success."[37] But if this is the case, if one can say that the form and manner in which knowledge is won by empirical means has this recursive character inscribed at its very heart, if it can be shown historically that "theories become clear and 'reasonable' only *after* incoherent parts of them have been used for a long time,"[38] then it also follows that in the concrete, practical process of the search for knowledge, explicit methodological rules are rather counterproductive, only causing additional confusion where they are supposed to produce clarity. Even a "law-and-order" science, which does have its place in Feyerabend's scientific universe—though not an exclusive one—must occasionally allow "anarchistic moves" if it is to remain on the agenda.[39] For, as Feyerabend later summarized: "Without ambiguity, no change, ever."[40]

The moment of contingency in the development of the sciences is emphasized by Feyerabend to a far higher degree than by any of the thinkers discussed so far. Contingency intervenes in the actual course of history in a way that lies beyond any critical-rational schematism. But Feyerabend was not ready to bind contingency into a developmental model, such as, for example, Kuhn or Toulmin did in their different versions of an evolutionary mechanism. And since objects of knowledge do not come into being following any general rule, Feyerabend saw it as superfluous to consider a methodology of the history of science in any detail. For history of science, and for its subjects, the principle was "do what you like." Yet for Feyerabend, the history of science takes over, for all that, as an inexhaustible reservoir of exemplary rule breaches against rationalism. It takes the place and becomes the heir of all analytic and rationalist philosophy of science.

5

The 1960s in France

A tradition of historical epistemology was already established in France in the 1930s, with the work of Gaston Bachelard, along with that of historian of chemistry Hélène Metzger and historian of mathematics Jean Cavaillès. After the war, this tradition was continued first of all by Bachelard himself, with a second series of epistemological texts. Bachelard's successor in the chair of history and philosophy of the sciences at the Sorbonne was the historian of biology and medicine Georges Canguilhem, who in turn had an influence that must not be underestimated on the group of French philosophers and historians prominent in the Paris debates around structuralism and poststructuralism in the 1960s. These included, among others, Michel Foucault, Louis Althusser, and Jacques Derrida. This chapter will examine some of their positions, insofar as they relate to basic questions of the history concerning science.

Canguilhem, who had studied not only philosophy but also medicine, focused on the history of medicine and the life sciences, whereas Bachelard had come to the history of science from physics, chemistry, and mathematics. If Bachelard had made a major contribution to historicizing the philosophy of science of his time, Canguilhem proceeded conversely from the history of science, and was concerned to give it an epistemological foundation with a focus on the history of scientific concepts. "This history," he wrote in an essay on the history of science in Bachelard's epistemological work, "can no longer be a collection of biographies, nor a tabulation of doctrines after the manner of a natural history. It must be a history of conceptual filiations. But this filiation possesses a discontinuity, just like Mendelian inheritance. The history of science must be as equally demanding, and as equally critical, as science itself."[1] Canguilhem represented a form of conceptual history that can also be understood as a history of the displacement of problems which must be reconstructed in their historic context.

Why history of science, and what for? For Canguilhem, the essential philosophical reason for taking the history of the sciences seriously was grounded in the fact that "a theory of knowledge with no reference to epistemology would be a meditation in the void, while epistemology with no reference to the history of the sciences would be a completely superfluous duplication of the science about which it pretends to have something to say."[2] If one really wants to know how scientific research functions, one has to look at the sciences in detail. This is best done, in the view of Canguilhem, following his Dutch colleague Eduard Dijksterhuis, by seeing the history of the sciences as itself an "epistemological laboratory." This is where the above-cited passage on the necessity of a history of science that is pursued as critically as the sciences themselves takes its precise meaning. An experimental epistemology sees "the relationship between the history of the sciences and the sciences whose history it is as just like . . . the relationship of the sciences to the objects whose sciences they are."[3] It is the task of the history of science to constitute a historical object sui generis, and to develop it further in constant debate and interaction with the historical material.

But how is this object to be grasped and understood in more detail? Canguilhem decidedly distances himself from two opposite positions that were put forward in the contemporary debate over the history of ideas: so-called externalism and internalism. Staunch externalists saw the history of the sciences as a derivative phenomenon, the explanation of which required reference to social and economic interests, and the technologies and ideologies that were bound up with them—interests that exhaustively determined science. The internalists, on the other hand, set themselves to what Canguilhem called a "completely superfluous duplication" of the sciences, simply following their course reflectively once more with the sciences' own conceptual resources. Both of these positions, which are counterposed here in a simplified version, thus fell short of reflection on the specificity of their object. The externalists, upon close inspection, lacked a special object called "science" to which a life of its own could be ascribed, while the internalists did not succeed in differentiating their own object from the objects of the science that they investigated.

Canguilhem separated three levels of objects. First, there is the natural object "outside of any discourse held about it." It functions vis-à-vis the possible scientific discourses like a kind of "pre-text." But it is not properly the object of the sciences, for this object constitutes a second level. "Nature is not in itself divided into scientific objects and phenomena. Science rather constitutes its object from the moment that it finds a method for building a theory from coherent propositions, which in turn is controlled by the concern to make out its flaws."[4] We are reminded here of Bachelard's dictum that the structure of scientific thought is "the awareness of its historical errors." The sciences submit their object to permanent change, yet they do not conceive of its essence as a historical one. On this second level, therefore, we still have not reached the object of the history of science, as this is once more sharply distinguished from the object of science itself. History of science has to identify and analyze the conditions under which the "secondary, nonnatural, cultural objects" of the sciences are being formed. The object of the history of science is therefore the particular "historicity of scientific discourse, insofar as it expresses a procedure that is normalized from within, but is punctuated by accidents, impeded or thwarted by obstructions, and interrupted by crises, that is, moments of judgment or truth."[5] The objects of the sciences are not treated as historical by the sciences themselves, yet from the perspective of the history of science the character of a historical process is inscribed into their very core.

The history of science thus constitutes a "specific domain in which the theoretical questions thrown up by scientific practice in its development find their place."[6] It was clear to Canguilhem that this domain was not a homogeneous space. On the one hand it was multiply partitioned, and the history that these particular partitions underwent was not reducible to a single homogeneous time. "The time of arrival of scientific truth, however, the time of veri-fication, has a different fluidity or viscosity for different disciplines in the same periods of general history."[7] On the other hand, the domain of history of science was not limited to scientific discourse in the narrower sense. Historians of science, in the context of their reconstructions, have to deal with science together with "nonscience, with ideology, and with political and social practice."[8] For the objects of the sciences are themselves the product of a cultural activity, and while their existence certainly maintains reference to the natural objects of the first order, they do not coincide with them.

The objects of knowledge are reflected in concepts, and by way of their changing meaning the historian is able to reconstruct trajectories. According to Canguilhem, it is only this last task that gives the history of science the rank of an activity that can itself make a claim to science, raising it above a mere inventory, the simple "natural history of a cultural object."[9] Canguilhem devoted his own work in the history of science to a conceptual history of this kind—from the concepts of the normal and the pathological in his medical dissertation; via the concept of reflex in the seventeenth and eighteenth centuries in his philosophical dissertation; to the biological concepts of the cell, the internal environment, and organic regulation, among others, in his later writings.

According to Canguilhem, historians must be aware of the fact that in their work they are constructing a temporal order of their own. One of the main problems that he raises in this context is the question of continuity and break in the development of the sciences. Here Canguilhem takes a cautious distance from Bachelard, at least as far as Bachelard's assumption of a radical break between everyday experience and scientific experiment is concerned. For Canguilhem, this transition is a smoother one. The sciences, moreover, as cultural formations of a particular kind, "breathe" as it were with a varying frequency, the dynamic of their conceptual replacement and transformation being sometimes slower and sometimes faster. But Canguilhem does not follow Kuhn's characterization of the structure of scientific revolutions either, at least not in the social and psychological dimension that Kuhn ascribes to the revolutionary gestalt switch.

Canguilhem seeks to make clear his own position with respect to the figure of the "precursor" in the history of science. For him, the precursor is the example of a "false historical object."[10] Both the precursor and the search for precursors are the result of a confusion between the object of science and the object of the history of science. "Strictly speaking," writes Canguilhem, "the history of science would lose all its meaning if there were precursors, since science would then only seemingly have a historical dimension."[11] The precursor is "not the agent of scientific progress," not an innovator *avant la lettre*, but rather a historiographic misconstruct.[12] Even in its faulty guise, however, it refers to the fact that a certain knowledge can in retrospect find a place that it did not have on its first appearance. With the notion of historical recurrence, Canguilhem (here closely following Bachelard) sought to develop the conception of something like a dialectical unity of continuity and change in the development of scientific knowledge, and to explain with examples such as the development of cell theory or the doctrine of biological inheritance how the objects of knowledge are historically constituted.

. . .

With Michel Foucault's archaeology of knowledge, French reflection on the historiography of science reached a new stage. Foucault had studied psychology and philosophy—his teachers included Louis Althusser, Jean Hyppolite, and Maurice Merleau-Ponty; in the early 1950s he was also much concerned with Heidegger, and first made his mark in the history of science with two major works on the history of psychology and medicine, *The History of Madness* and *The Birth of the Clinic*. Both books focused on the boundary between the natural and the human sciences. But it was only with *The Order of Things* in 1966 that Foucault attracted continuing attention beyond the French borders. In this "archaeology of the human sciences," he developed—with reference to economics, linguistics, and natural history of the eighteenth and nineteenth centuries—the concepts of "discourse formation" and "dispositive," which in *The Archaeology of Knowledge*, published in 1969, he then made into the object of a detailed historiographic reflection. Here Foucault claimed programmatically: "I cannot be satisfied until I have cut myself off from 'the history of ideas,' until I have shown in what way archaeological analysis differs from the descriptions of 'the history of ideas.'"[13] The traditional history of ideas, which Foucault set out to challenge, stood in his formulation under the three fatal stars of "genesis, continuity, totalization."[14] How an "archaeology" of knowledge undermines this triad is something we must examine more closely; for Foucault, as he never grew tired of stressing, it was above all a method of analysis "purged of all anthropologism."[15]

Foucault explicitly linked up here with the studies of Bachelard and Canguilhem, who had shown him that what characterized the history of the sciences was not homogeneity and continuity, but rather dispersion and breaks. He saw the underlying change that this meant in the form and manner in which history was seen as having come into being with Karl Marx in the later nineteenth century. For Foucault, this change was still far from complete, or at least not yet consistently taken up as the subject of reflection in the field of the history of knowledge—in contrast for instance with linguistics, psychoanalysis, and ethnology. "It is as if it was particularly difficult, in the history in which men retrace their own ideas and their own knowledge, to formulate a general theory of discontinuity, of series, of limits, unities, specific orders, and differentiated autonomies and dependences."[16] Genesis (with its other face of teleology), continuity, and totality were characteristic of a history conceived under the "founding function of the subject." It is against this triple conjuration of the subject that Foucault's effort was directed to put a stop to being afraid of the noise and proliferation of discourse, the "powerful anonymous murmuring of that discourse revolving on its own axis,"[17] and to conceive difference, decentration, the abandonment of coherence, in short "to conceive of the *Other* in the time of our own thought."[18]

There are four principles that distinguish an archaeology of knowledge from the traditional history of ideas. First and foremost, there is the unit of analysis, which is discourse. "Archaeology tries to define not the thoughts, representations, images, themes, preoccupations that are concealed or revealed in discourses; but those discourses themselves, those discourses as practices obeying certain rules."[19] It is, then, not a matter of ideal objects, or ideas, that could be peeled off from practiced discourses and their historical remnants in the sense of an appropriation of sources, but rather of the discursive conditions of possibility for producing certain things in speech and practice. It is a question of spontaneous or organized forms of "rémanence,"[20] models of persistence that do not present to the archaeologist more or less transparent documents for a decipherable ideality, but rather monuments whose own materiality and forms of resistance need to be addressed and understood.

Second, Foucault's archaeology sets its sights less on transitional zones than on what is typical of the strata in question. It "does not seek to rediscover the continuous, insensible transition that relates discourses, on a gentle slope, to what precedes them, surrounds them, or follows them." Rather it is interested in showing "in what way the set of rules that they put into operation is irreducible to any other."[21] Foucault is prepared to give this concentration on a "positivity" the title of structural analysis, but not as a way of counterposing it to historical analysis. The question for him is rather to undermine the traditional opposition of structure and genesis, and to turn this as it were on its head: history *is* structure.

Third, this archaeology's ordering principle and criterion of classification is not the finished work of a sovereign author. "The authority of the creative subject, as the *raison d'être* of an *oeuvre* and the principle of its unity, is quite alien to it." Foucault's archaeology rather defines "types of rules for discursive practices that run through individual *oeuvres*."[22] Such "running through" is to be taken quite literally here, as in seeking to track down rules of discourse Foucault is not setting out to define a new form of totality. What he is looking for is rather something like conditions of possibility that in the monuments take different material shape but can only be disclosed by incursion.

Fourth and last, "archaeology does not try to restore what has been thought, wished, aimed at, experienced, desired by men in the very moment at which they expressed it in discourse." It does not set out to uncover an original meaning that may have been buried in the course of time. It is not reducible to the authenticity of what is brought to speak. Rather it sees itself as "nothing more than a rewriting." It is, in other words, a construction and not an interpretation or a hermeneutics. If it brings something to light, it is a "discourse-object."[23]

In an archaeology of knowledge such as Foucault strove for, the focus of interest is not on a history of scientific disciplines. The use of the word *knowledge* in the title of his book was carefully chosen by Foucault. Discursive formations may include disciplines, but they need not in every case crystallize into disciplinary specialties. In his *Archaeology of Knowledge*, Foucault defined four thresholds for the formation of knowledge in the space of the discursive: the thresholds of positivity, of epistemologization, of scientificity, and of formalization. The threshold of positivity involves the separating off and staking out of a discursive field, which assumes a tendency to autonomy. Epistemologization applies within a discursive formation characterized by a particular positivity, for example through norms of coherence. Beyond the threshold of scientificity, statements within a formation must obey particular laws of construction. The threshold to formalization is finally crossed when the corresponding knowledge formation has made the transition to conceiving itself in terms of axiomatic assumptions.

Foucault's characterizations are somewhat unspecific in their formulation, but they are interesting from a historiographic point of view, as they enable him to associate previous approaches to the history of science, including his own, with a privileged gaze at particular thresholds. Traditional history of science, with its orientation to mathematical physics, preferred to operate above the threshold of formalization, and thus remained fixed on the narrow window of its very specific type of discourse, which it also marked out as normative. With the historical epistemology of Bachelard and Canguilhem, history of science was conducted at the threshold defined by scientificity. Bachelard's definition of the "epistemological break" between everyday knowledge and scientific knowledge pointed precisely to this threshold, according to Foucault. The archaeology of knowledge in turn directs its attention to the two thresholds of positivity and epistemologization. This is why Foucault describes it as an "analysis of the *episteme*."[24] It is the task of a future archaeology of discursive formations, conducted in the perspective of an emergence of epistemological figurations, including the sciences, to determine the mutual interrelations between these thresholds, which moreover in no way need to be crossed in the given sequence, and accordingly do not present a series of stages: "The distribution in time of these different thresholds, their succession, their possible coincidence (or lack of it), the way in which they may govern one another, or become implicated with one another, the conditions in which, in turn, they are established, constitute for archaeology one of its major domains of exploration."[25] It does not subsume knowledge under a theory of knowledge, but relates it to the "processes of a historical practice."[26]

What is decisive for archaeology of knowledge is that, as Foucault contends, it follows a "discursive practice/knowledge (*savoir*)/science axis," in contrast to traditional history of ideas, which moves along the "consciousness/knowledge (*connaissance*)/science axis." From this it follows that archaeology finds its point of reference, not in the *connaissance* of a subject, but in the *savoir* that is related to the structure of a practice, and in relation to which the subject in question does not appear as "titular," but essentially just as a temporary participant.[27] The archaeology of knowledge has recourse to deep structures, historically located rules of production which give an episteme its form, without being visible and transparent as such. With Foucault we are thus also dealing with a historical a priori, but one of a quite different form than that which we have seen with Husserl.

Foucault did not maintain, as is sometimes claimed, the disappearance of history and historicity altogether, but rather the evaporation of "that form of history that was secretly, but entirely related to the synthetic activity of the subject."[28] The history of ideas is itself a form of discourse, and Foucault inveighed against its "ponderous, awesome materiality" in *The Discourse on Language*,[29] his inaugural lecture at the Collège de France, which he gave a short time after publishing *The Archaeology of Knowledge*. He aimed at replacing it by a form of presentation that oriented itself on the structural sciences, whose object is not consciousness but rather ensembles of practices, such as economics, forms of communication, institutions, or rituals.

. . .

With that, to a certain extent, Foucault linked up with one of his mentors, Louis Althusser, even though the latter did not concern himself with history of science in any strict sense. Althusser arrived in France from Algeria before the war. He studied philosophy at the École Normale Supérieure, where he assumed a teaching post in 1948, after having spent the greater part of the war in a German labor camp. Althusser's reading of Marx, directed against both humanistic and orthodox interpretations, started out from the thesis that Marx had opened up a new continent of science with historical materialism, the continent of history. Althusser followed Marx in seeing the driving moment of history as the changing relationship between forces and relations of production, and hence saw its basic structure as a "process without a subject." In a late reflection on this initial position, he replied in an interview: "This is the negation of all teleology, whether rational, moral, political, or esthetic. I would add that this materialism is the materialism, not of a subject (whether God or the proletariat), but of a process—without a subject—which dominates the order of its development, with no assignable end."[30] This materialism can ultimately be defined only as an "aleatory" one, without which history cannot be conceived as "living history" in the sense of an "openness of the world to the event."[31] Althusser's interpretation of Marx could also be described as a structuralism of contingency.

Analogously, Althusser sought to conceive of scientific knowledge as an inconcludable process of production. As distinct from material production, however, while scientific work proceeds from a "real object," its products take the form of "objects of knowledge." Its practice is founded in the materiality of an experimental dispositive, the results of which, however, in their difference, as "thought-things" (*Gedankenkonkreta*, in Marx's expression), permanently need to be reported back to the real object. In his inaugural lecture, Althusser summarized this movement as follows: "This is the unending cycle of any knowledge, which only adds *its* knowledge to the real in order to render it back, and this cycle is only a cycle and consequently a living one, *if it reproduces itself*, for only the production of new knowledges keeps the old ones alive."[32] "Proof and test" are not externally applied standards, but rather "the product of definite and specific, material and theoretical dispositions and procedures, which are particular to each science."[33] There are thus two themes in Althusser's writings that Foucault took up, giving them his own stamp. First of all, the theme of a radical critique of the subject in the sense of a theoretical antihumanism, serving as the basis for an aleatory understanding of history; and second, the attempt to approach theory, including the sciences, from the standpoint of specific practices, and not in the form of concepts only.

. . .

The final writer to consider in this context is Jacques Derrida. It is well known that the natural sciences and their history play no role in his work—neither did they have a prominent place in Foucault's writings. Yet Derrida's work is intimately related to historical epistemology. Born like Althusser in Algeria, Derrida arrived in France in 1952, studying philosophy at the École Normale Supérieure and attending the seminars of Althusser and the young Foucault. In a talk from the late 1990s, Derrida had this to say about the time at which he was studying: "In the early 1950s, after phenomenology was introduced to France by Sartre and Merleau-Ponty, I saw the need to pose the question of science and epistemology, which neither Sartre nor Merleau-Ponty had done. And so I wrote my first essays on Husserl. These revolved around the questions of scientific objectivity and mathematics: Cavaillès, Tran Duc Thao, as well as the question of Marxism."[34]

It is the late Husserl, in fact—to whose phenomenological analyses both Sartre and Merleau-Ponty also referred—that best enables us to understand the path taken by Derrida. As we have seen, Husserl had opened up the space for an epistemology which posited itself in a historical perspective in a specific way. Derrida described it as follows: "In order to peel away this skin from the phenomenon, and differentiate it both from the reality of the thing and from the psychological texture of my own experience, an extraordinarily subtle operation was needed. It required recourse to the wilds of meaning, a particular sensibility in the conversion of the gaze."[35] In engaging with Husserl, Derrida located himself in the French tradition, situated between positivism and psychologism, that led from Bachelard, via Cavaillès and Canguilhem, to Foucault. It must not be forgotten that from 1960 to 1964, Derrida was assistant to Canguilhem at the Sorbonne.

The question is, to put it in Cavaillès' terms, how the movement of a "permanent revision of contents by deepening and extinction" can be appropriately conceived,[36] characterizing as it does the developmental process of the sciences, a process that takes place "in the wilds of meaning" and that is yet not devoid of rigor or rule. Proceeding from Husserl's remarks about the origin of geometry, Derrida hopes, as he notes, that on the one hand this effort "brings to light a new type or new profundity of historicity; on the other hand, and correlatively, it determines the new tools and original direction of historic reflection."[37] Derrida agrees with Husserl that "the historicity of ideal objectivities" of mathematics, "i.e. their *origin* and *tradition* (in the ambiguous sense of this word which includes both the movement of transmission and the perdurance of heritage)," obeys "different rules, which are neither the factual interconnections of empirical history, nor an ideal and ahistoric adding-on."[38] And at another place in *Edmund Husserl's "Origin of Geometry"* he adds: "If we take for granted the philosophical nonsense of a purely empirical history and the impotence of an ahistorical rationalism, then we realize the seriousness of what is at stake."[39]

The *Rückfrage*, as Derrida calls it after Husserl, is "the pure form of every historical experience."[40] It is this query that endows with meaning an event that would not otherwise be able to make its appearance and that as such would neither be accessible to a primary determination nor in a position to induce a discourse. As we have seen, however, Husserl's *Rückfrage* remained embedded in what he conceived as a "teleology of reason" under universal signs. Proceeding from Husserl, Derrida undertook the attempt to, as it were, turn the tables and transform the *Rückfrage* into a nonteleological form of iteration.

Husserl's *Origin of Geometry* contains just one short paragraph indicating the historical intervention of writing and its significance for the development of the sciences: "The importance of written, documenting linguistic expression is that it makes communications possible without immediate or mediate personal address; it is, so to speak, communication become virtual."[41] Derrida has stressed this indication of Husserl's, basing himself on it to develop a more exoterically oriented epistemology of the historical, one that operates in the space of "virtual communication" and that ascribes the sciences a place in a typology of forms of iteration. What this then involves is a knowledge of the *procedures* of obtaining knowledge. The space of phenomena and events that Derrida outlines and maintains—between the "reality of the thing" and the "psychological texture of experience"—thus becomes a place in which the *means* involved in the production of knowledge come into view. Parallel with Foucault's analysis of discourse, Derrida thus seeks to reflect on writing as such a means, whose instrumentality he aims at determining in the form of a historical logic of supplementarity. In recent decades history of science has filled this space with investigations on a large number of further media and devices of inscription that are characteristic of the development of the modern empirical sciences and that, to use Bachelard's concept once again, belong to the arsenal of their "phenomeno-techniques."

In his *Grammatology*, Derrida coined the term *historiality* for this iterative-recursive production of meaning in the irrevocable exteriorization of a generalized writing. Historiality goes beyond a mere chronology of events; at the same time, it remains this side of any teleology. The concept of "trace" is central to Derrida's thought about historiality. Writing has the character of a "trace." This concept also shows the distance that Derrida obtains in relation to Husserl, by no longer conceiving the movement of knowledge from an origin toward an end point (no matter how provisional), but rather in terms of its medium: "The trace is not only the disappearance of origin—within the discourse that we sustain and according to the path that we follow it means that the origin did not even disappear, that it was never constituted except reciprocally by a nonorigin, the trace, which thus becomes the origin of the origin."[42] It is only where something is envisaged as a trace that the origin can be conceived as what it is: not a starting point, but rather a construction that is by necessity belated.

Meaning, which for Husserl remained ultimately related to the illusion of an evidence of origin, for Derrida thus lies in displacement itself as a generator of meaning, that iterative process for which, in an essay of 1968 that followed his *Grammatology*, he used the wordplay of *différence/différance*. Here, meaning for Derrida is no longer a transcendental phenomenon but rather a diacritical one. The pun involved in his neologism consists in the fact that the difference in vowel is perceptible only in writing. In speech it disappears. Derrida characterizes *différance* as follows:

In the delineation of *différance* everything is strategic and adventurous. Strategic because no transcendent truth present outside the field of writing can govern theologically the totality of the field. Adventurous because this strategy is not a simple strategy in the sense that strategy orients tactics according to a final goal, a *telos* or theme of domination, a mastery and ultimate reappropriation of the development of the field. Finally, a strategy without finality, what might be called blind tactics, or empirical wandering if the value of empiricism did not itself acquire its entire meaning in its opposition to philosophical responsibility. If there is a certain wandering in the tracing of *différance*, it no more follows the lines of philosophical-logical discourse than that of its symmetrical and integral inverse, empirical-logical discourse. The concept of *play* keeps itself beyond this opposition, announcing, on the eve of philosophy and beyond it, the unity of chance and necessity in calculations without end.[43]

Derrida did not give these basic considerations concrete form in terms of a separate contribution to the history of science. Yet it is not hard to see that the "calculations without end" relate to a meditation on precisely the process of scientific research—"on the eve of philosophy and beyond it"—in which the ultimate task is to engender the new that by its nature is unprecedented and unpredictable.

6

Recent Developments

This brief concluding chapter will focus on two exponents of more recent theoretically motivated history and philosophy of science. As we shall see, Ian Hacking and Bruno Latour both exemplify, if in different forms, the return to a theme that was to a large extent anathema for French poststructuralism discussed in the previous chapter: the question of a historical anthropology of the sciences.

The Canadian philosopher of science Ian Hacking did his doctorate at Cambridge. He went on to teach in Toronto and later at the Collège de France in Paris. In his *Representing and Intervening*, he argued that the publication of Thomas Kuhn's *Structure of Scientific Revolutions* had introduced a crisis of rationality in the understanding of the sciences, and drawn a definite line under the era of logical positivism and critical rationalism. According to Hacking, Kuhn's main theses were as follows: "There is no sharp distinction between observation and theory.—Science is not cumulative.—A live science does not have a tight deductive structure.—Living scientific concepts are not particularly precise.—Methodological unity of science is false . . . —The sciences themselves are disunified. . . . —The context of justification cannot be separated from the context of discovery.—Science is in time, and is essentially historical."[1]

After what has been shown so far, it is legitimate to doubt whether responsibility for all this is solely Kuhn's. We might rather assume that Hacking is summarizing here a protracted effort that took various forms over a good half-century, despite the temporary philosophical dominance of logical positivism, and that Kuhn's programmatic text bundled together. Hacking himself took the crisis of rationality in the image of science that he diagnosed as an opportunity to shift perspective from a theory bias in the characterization of the sciences that Kuhn had not challenged, and focused on scientific practice, as indicated in his programmatic title *Representing and Intervening*. Not dissimilar to Foucault in this respect—although, as we shall go on to see, in the context of a completely different tradition—he was determined to get away from the "history of ideas, or intellectual history," even in the punctuated and noncumulative form that Kuhn had proposed. In Hacking's words, "There is a simpler, more old-fashioned concept of history, as history not of what we think but of what we do."[2]

Hacking dedicated his philosophy of science to the promotion of a corresponding practical turn in the historiography of science, a task that was carried on by others in the following decades. Hacking started from the premise that theories were not the fundamental entities for understanding the phenomenon of the sciences and their dynamic. Theories were rather always already embedded in the most varied contexts of practice and experiment. Modern science did not simply consist in representation, but always already had an interventionary character. "Natural science since the seventeenth century has been the adventure of the interlocking of representing and intervening. It is time that philosophy caught up to three centuries of our own past."[3] Hacking accordingly laid value on characterizing the experimental aspect of acquiring knowledge in its multifarious modes and shades, and above all on elaborating its autonomous character. The coupling of the theoretical and experimental aspects in the acquisition of knowledge, for Hacking, could take widely different forms, and was in no way reducible to the imagined ideal of a hypothetico-deductive schematism, in which experiment is limited to testing a concrete proposition derived from theory.

Experiments can generate new phenomena that were previously inaccessible, that prove stable and yet cannot be brought into connection with any kind of existing theory. They can lead to empirical regularities and thus structure fields of experience, without a comprehensive theoretical explanation of such regularity being to hand. Theoretical solutions can even under certain circumstances be delayed a long while, arising eventually from contexts that did not originally belong to the field of experimentation in question, and yet prove fruitful in the new context. Hacking saw it as an advance to recognize the rich table of experimental interventions as given, even if it could not be reduced to a general rule or be fully integrated at a particular point in time. He summed up this position in the later widely followed advice: "Think about practice, not theory."[4] "Experimentation"—in Hacking's lapidary formulation—"has a life of its own."[5]

But apart from his emphasis on the experimental or interventionary character of modern science, Hacking also gives the theoretical aspect of knowledge of the world—its representational character—a new foundation. He couches it in the form of an anthropological myth of origin and describes it as a philosophical anthropology. Our concept of "reality," according to Hacking's basic thesis, *is just a byproduct of an anthropological fact*."[6] This fact rests on the situation that to represent is part of human nature, indeed that "man" can even be defined as a representational animal. It is not rationality that distinguishes man, not language or upright gait, nor even the ability to produce tools—all themes that have marked Western philosophical anthropology—but rather the ability to represent.

Representation is originary, and in the beginning even itself a material activity, a practice, which consists in creating objects that resemble other things and from the beginning are themselves things in a public space. Representation is therefore primary, and precedes the formation of a concept of reality, to which it refers: "The first peculiarly human invention is representation. Once there is a practice of representing, a second-order concept follows in train. This is the concept of reality, a concept which has content only when there are first-order representations."[7] In this way Hacking turns the question round. It is not that we need a concept of reality in order to identify representations as likenesses, but rather we need representation as a practice in order to give conceptual expression to "the real as an attribute of representations."[8]

At this stage, the concept of the real remains unproblematic and directly tied to experience. A problem only arises when alternative systems of representation are available, and when we have to decide between them. "We make public representations, form the concept of reality, and, as systems of representation multiply, we become skeptics and form the idea of mere appearance."[9] The specifically modern form of the problem derives from the fact that serious alternative representations of mechanics—as Heinrich Hertz, for example, presented them in the introduction to his

Principles of Mechanics—and non-Euclidean geometries came into use in the course of the late nineteenth century. They played, as we have seen, a decisive role in the historicization of the concept of science in the early twentieth century. Its basis was the emergence of alternative practices of scientific representation that came to be used in relevant ways—both theoretically and experimentally.

. . .

In 1979, in the Greek town of Delphi, Hacking sketched out his philosophical anthropology, using it to give the old problem of representation and reality a new foundation, one designed to allow a new conception of scientific practice in general. The perspective that he took again brought in the subject that had been so persistently questioned in the French structuralist and poststructuralist tradition. Yet the price of this was to attribute to a human "essence" what Heidegger, for example, had seen as specific to the modern age, namely the "conquest of the world as picture," and thus to an episteme with a deep historic anchorage.[10]

In the same year—which also saw the first English translation of Ludwik Fleck's *Genesis and Development of a Scientific Fact*—Bruno Latour and Steve Woolgar published their ethnological study on the laboratory of the endocrinologist Roger Guillemin at the Salk Institute in California. The title of their book was *Laboratory Life: The Social Construction of Scientific Facts*. Here Latour—who had studied philosophy and trained as an anthropologist, conducting fieldwork in Africa—applied the principle of field study to analyzing science and technology in the making. He thus opened up a new access to scientific practice, which previously—even in the case of Fleck—had been pursued only in the form of a reflection on the research process by practicing scientists themselves.

In *Laboratory Life*, the problem of representation in the depiction of experimental phenomena was laid out as a process of "inscription."[11] Latour and Woolgar paid particular attention to the laboratory procedures of transformation and translation that lead, by the application of an arsenal of instruments, and via a shorter or longer series of intermediate stages, from a piece of matter to an inscription, in which form it eventually enters the circuit of scientific communication. In this process, as Latour later showed in his "photo-philosophical" essay on the "pedology thread" of

Boa Vista, "we never detect the rupture between things and signs, and we never face the imposition of arbitrary and discrete signs on shapeless and continuous matter."[12] It is in the space of this chain of intermediate links that the concrete work of the sciences on the phenomena they are interested in takes place. It is at the same time the space of construction of scientific facts. As with Hacking, the problem of reality, or what Latour calls the "reference," only arises as such in the course of representation itself. "It seems that reference is not simply the act of pointing or a way of keeping, on the outside, some material guarantee or the truth of a statement: rather it is our way of keeping something *constant* through a series of transformations."[13]

If the anthropologist Latour's microscopic examination was originally directed at what scientists and technologists actually do in their work, how they proceed when they experiment, how they handle and transform their materials, in the course of his later work he embedded this method in an attempt to sketch out a "symmetrical anthropology."[14] On its premises, the givens of our world are viewed from a perspective that fundamentally puts in question the separation between the natural, the social, and the discursive aspects of all the circulating objects that make up our modern world, including the separation between natural scientific, social, or cultural-historical reflection on the matters in question. Our world is pervaded by "sociotechnological networks," which are "*simultaneously real, like nature, narrated, like discourse, and collective, like society.*"[15] The paradox of modernity, for Latour, is that the efforts it has undertaken to tidy up the world by separating it into natural, social, and discursive phenomena, have produced as a side-effect a proliferation of mixed creatures whose hybrid character can no longer be concealed in the world of today. There are now, we could paraphrase Latour, no longer any significant, really serious problems that could be solved in a satisfactory way with the resources of *a single one* of these tidied-up spheres.

This means, however, that the sciences and technologies that for the last three hundred years have been the motor of this proliferation of hybrids can no longer be considered simply from one of these perspectives either. "For twenty years or so," Latour wrote in 1991 in *We Have Never Been Modern*, "my friends and I have been studying these strange situations that the intellectual culture in which we live does not know how to categorize. For lack of better terms, we call ourselves sociologists, historians, economists, political scientists, philosophers or anthropologists. But to these venerable disciplinary labels we always add a qualifier: 'of science and technology.'"[16] And just as, in the early twentieth century, the event of the First World War effected an irrevocable break in the positivist belief in the unstoppable progress that science and technology would bring to humanity, so at the end of the century, according to Latour, the double event of the collapse of the socialist camp and the threat of a global ecological crisis point to a limit from which it will be possible both to understand the previous adventure of modernity in a new way and to newly conceive a nonmodern future.

The means for this change of mind, according to Latour, can only be provided by anthropology, which has long been concerned with "the seamless fabric of what I shall call 'nature-culture.'"[17] It is just that up to now, in the self-understanding of our modern age, it is only *other* people who have been subjected to this mode of anthropological attention and are to be understood accordingly, whereas *we* have devoted ourselves to the separation of these regions and hence not included our cultures in the same symmetrical-anthropological view. If the modern world is to be made into an object of anthropology, then "the very definition of the modern world has to be altered."[18] And this is precisely Latour's intention, no more and no less: "And what if we had never been modern? Comparative anthropology would then be possible. The networks would have a place of their own."[19]

Finally, the separating work of modernity, which according to Latour has created two "ontological zones"—human nature on the one hand and nonhuman things on the other—conceals the fact that there is a second and no less active modern ensemble of practice, which by "translation" creates ever new and previously nonexisting "hybrids of nature and culture."[20] We observe here, moreover, as a macrostructure, the appearance of the same figure that Latour had introduced in his microdescription of scientific practice with its mixed beings of "things/signs." Modernity number one separates by way of critique, while the second creates networks. Only in retrospect do we become aware that "the two sets of practices have always already been at work in the historical period that is ending,"[21] and that the project of critique only presents one side of it. Latour ascribes the task of sounding out new options for the following century with a concomitant new understanding of modernity's past neither to the natural sciences, nor to sociology, nor again to philosophy, but to studies of science conducted in terms of a symmetrical anthropology. These are, in his view, destined to appear as the heir to the philosophy, history, and sociology of science of the twentieth century.

. . .

Here we must raise the question as to how Hacking's and Latour's anthropological perspectives on the sciences and their history are situated with respect to the radical critique of anthropocentrism that we encountered in the tradition of the attempts at historicizing the scientific enterprise and its dynamic from Bachelard to Foucault. First of all, it is apparent that Hacking's and Latour's appeals to anthropology do not coincide; they offer two different options. Hacking grounds the option of a scientific grip on the world in ontology, and in a quite traditional way in a faculty of nature, the nature of man as a representing being, whose defining practice has produced a stream of knowledge in the course of human history—"people make likenesses."[22] Latour, on the other hand, sees anthropology rather as a procedure, a transdisciplinary possibility of access, that can break the modern claim of the sciences to a proud autonomy. His aim, and here he certainly follows Foucault, is to grasp the sciences' forms of thinking and construction of objects not as privileged but rather as always already embedded in a broader context of life, in which they have to be situated—and in which they also will be pursued in the future.

Common to both, however—and they are of one mind here, linking up with the tradition of historical epistemology despite all the criticism from Latour's side or neglect from Hacking's—is that they direct special attention to the *practices of obtaining knowledge.* Both give broad room for representation of the material side of the sciences, and, each in their own way, understand the step from the dynamic of theory to the dynamic of practice as opening a field of historical inquiry that obeys rules which cannot be derived from the conscious mind alone. Hacking stresses, coming as he does from the tradition of Anglo-Saxon pragmatism, that "representation" is from the very start, as he expresses it, "something public," something that takes place on a terrain that is socially constituted or that even underlies such a constitution.[23] For him every cultural tradition of productive mastery of life is based on a principle of similitude. Latour, in his network theory, endows things, as irreducible hybrids of nature and culture, with a central agent function, to the extent that they are ontologically equated to human agents in their actions. Thus, in his own manner he seeks to structure this world of intermediates in such a way that the history of knowledge and the sciences can find a place in it as one of many ongoing agonisms between humans and things.

A considerable part of historical studies of science over the past three decades has increasingly devoted itself to the question as to how scientific controversies, which have to be examined under concrete historical circumstances, in the context of given theoretical assumptions and with the scientific means available, can for all that be brought to a *closure*—as provisional as such closure might be. Epistemologically, this was the overarching theme of the Edinburgh sociology of knowledge school, whose protagonists, among them David Bloor, Barry Barnes, and Harry Collins, prescribed a "strong program" for the investigation of science. It demanded subjecting successful and failed scientific programs to a symmetrical factor analysis. In terms of the history of science, a whole group of historians in the wake of this program, including Simon Schaffer, Norton Wise, and Peter Galison, directed the focus of their investigations to the pragmatic, social, cultural, and not last rhetorical processes of normalization, standardization, and metrologization that could be observed in scientific practice. They set out to understand these processes of closure above all in terms of the technical systems that became, in particular in the course of the nineteenth century, important contextual conditions for further scientific development.

The complementary problem—how the processes of *opening up* of new scientific territory take place—remained virulent to some extent, as we have seen, for authors such as Latour and Hacking, and was particularly taken up in Andrew Pickering's concept of the "mangle of practice." My own work on experimental systems and epistemic things also belongs here.[24] At the same time, similar questions were tackled, even if with different means, and proceeding rather from biographical research trajectories, by those historians of science—Mirko Grmek and Frederic Holmes among them—who sought to reconstruct the concrete course of particular scientific paths of discovery from laboratory protocols and documents. Their aim was to illuminate the laboratory process in its creative dimension, not on the basis of its finished products, but rather in terms of the unpredictability of its results. These multifaceted studies, which cannot be pursued in further detail here, have all in one way or another shaped the narrative structures of history of science and enriched its historiographic practices with new considerations; in particular, however, their case studies have stimulated a lasting orientation for which the slogan of a "practical turn" in the study of science and its history has become current.

Conclusion

We have reached the end of this journey, which has led over a century of reflection on the sciences, their constitution, and their changes. It began with the idea of a kind of mimicry, the idea that the historical pursuit of science would follow the inductive course of the sciences, purged of its accidental hesitations. Via a series of shifts in the historical understanding of the relationship between science and technology, it led to the opening up of a field that took shape, not least in the debate with phenomenology after the First World War, and finally flowed into the quest for a new definition of the age of modernity at the end of the Cold War. What began as epistemological reflection emerging at the margins of classical mechanics opened out into different approaches and attempts at a genuine historical epistemology. It sought, steering its way between the poles of an empirically underpinned historicism based on the causal linkage of facts and a traditional, anthropologically motivated rationalism that privileged the consciousness of the knowing subject, to reveal the specific life of the sciences and their development.

In the course of time, historical reflection on epistemology began to merge with epistemological reflection on the history of science. It is no accident, seen from this perspective, that means and media have moved center stage—gradually but increasingly—in a comprehensive analysis of scientific practices in all their discursive and material dimensions. If it is ultimately from this shift that the question of a historical anthropology of the sciences has been newly raised, the latter should not be misconstrued as a return of anthropocentrism, either in its empiricist-decisionist variant or in its rationalist-creativist one. It should rather be read as an attempt, in the context of a thoroughly altered system of coordinates of the growth of science, no longer defined in Cartesian terms, to newly assess the role of human actors and their ever changing position in a network that embraces them and yet allows them to remain decentered.

The building blocks of a new, genuinely historical-epistemological discourse, which initially came rather from the margins of established scientific disciplines, were likewise introduced by outsiders in a discussion that was at the beginning still carried on in terms of academic philosophy. More than a few of these figures came out of the sciences themselves. As could also be observed, we are not faced from the start with a continuous discourse handed down from one scholar to the next. The twentieth century, with its major political events and its legacy of national traditions, was too riven to allow such continuity. Moreover, the intellectual migration forced by National Socialism also tore up existing traditions, especially in the German language zone. The dislocations and international reshufflings this brought about have still not been worked through in terms of a history of philosophy of the twentieth century.

And yet, as the positions presented in these brief portraits show, there was a persistence of a set of problems, which time and again arose from different perspectives and in different contexts. These problems were raised and re-actualized repeatedly by the developmental dynamic of the twentieth-century sciences themselves. If we wanted to seek a continuity, it would be the continuity of changes and breaks that the sciences underwent in this century. Correspondingly, it can be stated that at the end of the century there is no longer any epistemology fruitfully intervening in discussion of philosophical questions of the sciences that is not permeated by historical questions. The idea of a linear development of knowledge, continuous and cumulative, from a teleological perspective, has gone, along with the idea of a unitary science that would embrace everything, centered firmly in physics. In its place, however, as the preceding presentation has shown, we do not have a new prevailing and compelling model. The space of historical epistemology has itself become plural in keeping with the course of its development. Perhaps it is a lesson learned from the pluralization process of the sciences in the twentieth century that such unity is not needed in order to advance. Historical epistemology has its own permanent laboratory in the past and future history of the sciences.

REFERENCE MATTER

Notes

INTRODUCTION

1. In Britain it was the statistician Karl Pearson's *Grammar of Science* that held this place, and in the United States various variants of pragmatism. Their views are not followed in the present essay.

CHAPTER 1: FIN DE SIÈCLE

1. Du Bois-Reymond 1912a, p. 435.
2. Duhem 1991, p. 268.
3. Du Bois-Reymond 1912b, p. 447.
4. Mach 1959, p. 313.
5. Mach 1960, p. 559.
6. Ibid., p. 316.
7. Ibid., p. 579.
8. Ibid., p. 581.
9. Ibid., p. 586.
10. Ibid., p. 612.
11. Boutroux 1916, pp. 12–13.
12. Ibid., p. 13.
13. Boutroux 1914, p. 165.
14. Boutroux 1916, pp. 165–66.
15. Boutroux 1914, p. 214.
16. Ibid., p. 217.
17. Dilthey 1988, p. 296.
18. Ibid., pp. 296, 300.
19. Dilthey 1960, pp. 3, 6, 11.
20. Poincaré 1952, p. 146.
21. Poincaré 1958, p. 14.
22. Poincaré 1905, preface.
23. Poincaré 1958, p. 139.
24. Dilthey 1960, p. 13.

25. Neurath 1973, p. 101.
26. Neurath 1983, p. 13.
27. Ibid.
28. Neurath 1973, pp. 102 ff.
29. Neurath 1983, p. 14.
30. Neurath 1973, p. 109.
31. Ibid., p. 101.
32. Ibid.
33. Ibid., p. 105.
34. Ibid., p. 102.
35. Ibid., p. 112.
36. Neurath 1983, p. 28.
37. Neurath 1973, p. 112.
38. Neurath 1983, p. 30.
39. Ibid., p. 31.

CHAPTER 2: BETWEEN THE WARS—I

1. Bachelard 1984, pp. 2–3.
2. Ibid., pp. 3–4; emphasis in the original.
3. Ibid., p. 5.
4. Bachelard 2002, p. 7.
5. Bachelard 1984, p. 7.
6. Ibid., p. 6.
7. Ibid., p. 10.
8. Ibid., p. 13.
9. Ibid.
10. Wind 2001, p. 10.
11. Bachelard 1984, p. 12.
12. Ibid.
13. This term appeared in 1932 in the title of Bachelard's book *Le pluralisme cohérent de la chimie moderne*.
14. Bachelard 1984, p. 8.
15. Ibid., p. 12.
16. Bachelard 1949, p. 133.
17. Bachelard 1968, p. 27.
18. Bachelard 1984, p. 172.
19. Bachelard 1951, p. 25.
20. Bachelard 1984, pp. 173 ff.
21. Ibid., p. 160.
22. Bachelard 1972, p. 39.
23. Bachelard 1951, p. 21, cited after de Broglie 1947, p. 9.

24. Bachelard 1951, pp. 143 ff.
25. Fleck 1983, p. 46. All translations are by the translator of this volume unless otherwise indicated.
26. Ibid.
27. Ibid., p. 47.
28. Fleck 1979, p. 21.
29. Fleck 1983, p. 54.
30. Ibid., p. 55.
31. Ibid., pp. 48 ff.
32. Ibid., p. 50.
33. Ibid., p. 53.
34. Ibid., p. 48.
35. Ibid., p. 51.
36. Ibid., p. 54.
37. Fleck 1979, p. 99; emphasis in the original.
38. Ibid., p. 92.
39. Ibid.
40. Ibid., p. 89.
41. Ibid., p. 86.
42. Ibid.; emphasis in the original.
43. Ibid., p. 95.
44. Ibid., p. 69.
45. Cf. Rheinberger 1997.
46. Fleck 1979, p. 78.
47. Ibid., p. 177, note 4.
48. Ibid., p. 39; emphasis in the original.
49. Ibid., p. 42.
50. Ibid., p. 78.
51. Ibid., p. 31 of the German original only.
52. Ibid., p. 93 and *passim*.

CHAPTER 3: BETWEEN THE WARS—II

1. Popper 1968, pp. xvii, 3.
2. Ibid., p. 30; emphasis in the original.
3. Ibid., p. 37; emphasis in the original.
4. Ibid., p. 38.
5. Ibid., pp. 41 ff.
6. Ibid., p. 41; emphasis in the original.
7. Bachelard 1984, p. 172.
8. Popper 1968, p. xix; emphasis in the original.

9. Popper 1972, p. 145.
10. Ibid., pp. 145 ff.
11. Ibid., p. 112.
12. Husserl 1970a, p. 9.
13. Ibid., p. 14.
14. Husserl 1970b, p. 318.
15. Husserl 1970a, p. 6.
16. Husserl 1993, p. 35; translation by the author.
17. Ibid., p. 36.
18. Husserl 1970b, p. 318.
19. Ibid., pp. 342 ff.
20. Husserl 1978, p. 159.
21. Ibid.
22. Ibid., pp. 166, 171.
23. Ibid., p. 168.
24. Ibid., p. 159; emphasis in the original.
25. Ibid., p. 175.
26. Ibid.
27. Ibid., p. 171; emphasis in the original.
28. Ibid., p. 176.
29. Ibid., p. 180.
30. Heidegger 1977, p. 118.
31. Ibid., p. 119; translation modified.
32. Ibid., p. 123.
33. Ibid., p. 121.
34. Ibid., p. 123.
35. Ibid., p. 126.
36. Ibid., p. 124.
37. Ibid.
38. Ibid., p. 127.
39. Ibid., pp. 129–30.
40. Bachelard 1984, p. 6.
41. Bachelard 1949.
42. Heidegger 1977, p. 134.
43. Cassirer 1957, p. 447.
44. Ibid., p. 453.
45. Ibid., p. 459.
46. Cassirer 1950, pp. 17–19.
47. Ibid., pp. 18–19.
48. Cassirer 1957, p. 449.
49. Ibid., pp. 475 ff.

50. Cassirer 1961, p. 76.
51. Ibid., p. 81.
52. Ibid., p. 98.
53. Ibid., p. 145.
54. Husserl 1978, p. 176.
55. Cassirer 1961, p. 164.
56. Ibid., p. 172.

CHAPTER 4: AFTER 1945

1. Koyré 1957, p. 2.
2. Koyré 1966a, p. 74.
3. Ibid., pp. 73, 80.
4. Koyré 1966b, p. 191.
5. Ibid., p. 178.
6. Ibid., p. 179.
7. Butterfield 1965, pp. 11, 16.
8. Kuhn 1962, p. 1.
9. Ibid., p. 2.
10. Ibid., p. 3.
11. Cf. Kubler 1962.
12. Kuhn 1962, p. 8.
13. Ibid., p. 9.
14. Ibid., p. 11.
15. Ibid., p. 169.
16. Ibid., pp. 167–68.
17. Ibid., p. 170.
18. Ibid., p. 168.
19. Ibid., pp. 170–71.
20. Kuhn 1992, p. 14.
21. Toulmin 1961, p. 99.
22. Ibid., p. 100.
23. Ibid.
24. Ibid., p. 101.
25. Ibid., p. 110.
26. Ibid.
27. Toulmin 1972, p. 204.
28. Ibid., p. 232.
29. Ibid., p. 252.
30. Feyerabend 1975, p. 18.
31. Ibid., p. 18, note.

32. Feyerabend 1976, p. 34, note 14.
33. Feyerabend 1975, p. 23.
34. Ibid., p. 24.
35. Ibid., p. 26; emphasis in the original.
36. Bachelard 2001, p. 24; emphasis in the original.
37. Feyerabend 1975, p. 26.
38. Ibid.
39. Ibid., p. 27.
40. Feyerabend 1995, p. 179.

CHAPTER 5: THE 1960S IN FRANCE

1. Canguilhem 1968a, p. 184.
2. Canguilhem 1968b, pp. 11–12.
3. Ibid., p. 12.
4. Ibid., pp. 16–17.
5. Ibid., p. 17.
6. Ibid., p. 19.
7. Ibid.; "veri-fication" in the original.
8. Ibid., p. 18.
9. Ibid.
10. Ibid., p. 22.
11. Ibid., pp. 20–21.
12. Ibid., p. 23.
13. Foucault 1972, p. 136.
14. Ibid., p. 138.
15. Ibid., p. 16.
16. Ibid., p. 12.
17. Foucault 1994, p. 594.
18. Foucault 1972, p. 12; emphasis in the original.
19. Ibid., p. 138.
20. Ibid., p. 123.
21. Ibid., p. 139.
22. Ibid.
23. Ibid., pp. 139–40.
24. Ibid., p. 191.
25. Ibid., p. 187.
26. Ibid., p. 192.
27. Ibid., p. 183.
28. Ibid., p. 14.
29. Ibid., p. 216.

30. Althusser 2006, p. 260.
31. Ibid., pp. 264 ff.
32. Althusser 1977, p. 158; emphasis in the original.
33. Ibid., p. 132.
34. Derrida 1999, p. 20.
35. Ibid., p. 76.
36. Cavaillès 1947, p. 78.
37. Derrida 1989, p. 26.
38. Ibid; emphasis in the original.
39. Ibid., p. 51.
40. Ibid., pp. 50–51.
41. Husserl 1978, p. 164.
42. Derrida 1974, p. 61.
43. Derrida 1982, p. 7; emphasis in the original.

CHAPTER 6: RECENT DEVELOPMENTS

1. Hacking 1983, p. 6.
2. Ibid., p. 17.
3. Ibid., p. 146.
4. Ibid., p. 274.
5. Ibid., p. 150.
6. Ibid., p. 131; emphasis in the original.
7. Ibid., p. 136.
8. Ibid.
9. Ibid., p. 142.
10. Heidegger 1977, p. 134.
11. Cf. Latour and Woolgar 1979, *passim*.
12. Latour 1999, p. 56.
13. Ibid., p. 58.
14. Latour 1993.
15. Ibid., p. 6; emphasis in the original.
16. Ibid., p. 3.
17. Ibid., p. 7.
18. Ibid.
19. Ibid., p. 10.
20. Ibid.
21. Ibid., p. 11.
22. Hacking 1983, p. 132.
23. Ibid.
24. Rheinberger 1997.

Bibliography

Althusser, Louis. 1977. "Est-il simple d'être marxiste en philosophie?" In *Positions*. Paris: Éditions Sociales, 1976, pp. 127–72.

———. 2006. *Philosophy of the Encounter*. Trans. G. M. Goshgarian. London: Verso.

Bachelard, Gaston. 1928. *Essai sur la connaissance approchée*. Paris: Vrin.

———. 1932. *Le pluralisme cohérent de la chimie moderne*. Paris: Vrin.

———. 1949. *Le rationalisme appliqué*. Paris: Presses Universitaires de France.

———. 1951. *L'activité rationaliste de la physique contemporaine*. Paris: Presses Universitaires de France.

———. 1968. *The Philosophy of No*. Trans. G. C. Waterston. New York: Orion. [*La philosophie du non: Essai d'une philosophie du nouvel esprit scientifique*. Paris: Presses Universitaires de France, 1940.]

———. 1972. "Le problème philosophique des méthodes scientifiques" [1951]. In *L'engagement rationaliste*. Paris: Presses Universitaires de France, pp. 35–44.

———. 1984. *The New Scientific Spirit*. Trans. Arthur Goldhammer. Boston: Beacon. [*Le nouvel esprit scientifique*. Paris: Presses Universitaires de France, 1934.]

———. 2001. *The Formation of the Scientific Mind*. Trans. Mary McAllester Jones. Manchester, England: Clinamen. [*La formation de l'esprit scientifique: Contribution à une psychanalyse de la connaissance objective*. Paris: Vrin, 1938.]

Boutroux, Émile, 1914. *Natural Law in Science and Philosophy*. Trans. Fred Rothwell. London: David Nutt. [*De l'idée de loi naturelle dans la science et la philosophie contemporaines*. Course taught at the Sorbonne, 1892–93. Paris: Lecène, Oudin, 1895.]

———. 1916. *The Contingency of the Laws of Nature*. Trans. Fred Rothwell. Chicago: Open Court. [*De la contingence de la nature*. Paris: Germer Baillière, 1874.]

Broglie, Louis de. 1947. *Physique et microphysique*. Paris: Albin Michel.

Butterfield, Herbert. 1965. *The Whig Interpretation of History* [1931]. New York: W. W. Norton.

Canguilhem, Georges. 1968a. "L'histoire des sciences dans l'oeuvre épistémologique de Gaston Bachelard." In *Études d'histoire et de philosophie des sciences*. Paris: Vrin, pp. 173–86.

———. 1968b. "L'objet de l'histoire des sciences." In *Études d'histoire et de philosophie des sciences*. Paris: Vrin, pp. 9–23.

Cassirer, Ernst, 1950. *The Problem of Knowledge: Philosophy, Science, and History Since Hegel.* Trans. William H. Woglom and Charles W. Hendel. New Haven, Conn.: Yale University Press. [*Das Erkenntnisproblem in der Philosophie und Wissenschaft der neueren Zeit: Von Hegels Tod bis zur Gegenwart, 1832–1932*. Stuttgart: Kohlhammer, 1957.]

———. 1957. *The Philosophy of Symbolic Forms.* Vol. 3, *The Phenomenology of Knowledge*. Trans. Ralph Manheim. Oxford: Oxford University Press. [*Philosophie der symbolischen Formen*. Vol. 3 (1929). Hamburg: Felix Meiner, 2000.]

———. 1961. *The Logic of the Humanities.* Trans. Clarence Smith Howe. New Haven, Conn.: Yale University Press. [*Zur Logik der Kulturwissenschaften. Fünf Studien* (1942). Darmstadt: Wissenschaftliche Buchgesellschaft, 1961.]

Cavaillès, Jean. 1947. *Sur la logique et la théorie de la science*. Paris: Presses Universitaires de France.

Derrida, Jacques. 1974. *Of Grammatology*. Trans. Gayatri Chakravorty Spivak. Baltimore: Johns Hopkins University Press. [*De la grammatologie*. Paris: Minuit, 1967.]

———. 1982. "Différance." In *Margins of Philosophy*. Trans. Alan Bass. Brighton, England: Harvester. ["La différance." In *Marges de la philosophie*. Paris: Minuit, 1972.]

———. 1989. *Edmund Husserl's "Origin of Geometry": An Introduction*. Trans. John P. Leavey, Jr. Lincoln: University of Nebraska Press. [*Introduction à Edmund Husserl: "L'origine de la géométrie."* Paris: Presses Universitaires de France, 1962.]

———. 1999. *Sur parole*. Paris: Éditions de l'Aube.

Dilthey, Wilhelm. 1960. *Das geschichtliche Bewusstsein und die Weltanschauungen* [1911]. *Gesammelte Schriften*, vol. 8. Stuttgart: Teubner.

———. 1988. *Introduction to the Human Sciences*. Trans. Ramon J. Betanzos. Detroit: Wayne State University Press [*Einleitung in die Geisteswissenschaften* (1883). *Gesammelte Schriften*, vol. 1. Stuttgart: Teubner, 1922.]

Du Bois-Reymond, Emil. 1912a. "Über Geschichte der Wissenschaft" [1872]. In *Reden von Emil Du Bois-Reymond in zwei Bänden: Mit einer Gedächtnisrede von Julius Rosenthal*, ed. Estelle Du Bois-Reymond. Leipzig: Veit, vol. 1, pp. 431–40.

———. 1912b. "Über die Grenzen des Naturerkennens" [1872]. In *Reden von Emil*

Du Bois-Reymond in zwei Bänden: Mit einer Gedächtnisrede von Julius Rosenthal, ed. Estelle Du Bois-Reymond. Leipzig: Veit, vol. 1, pp. 441–73.

Duhem, Pierre. 1991. *The Aim and Structure of Physical Theory.* Trans. Philip P. Wiener. Princeton, N.J.: Princeton University Press. [*La théorie physique: Son objet, sa structure.* Paris: Chevalier and Rivière, 1906.]

Feyerabend, Paul. 1975. *Against Method: Outline of an Anarchistic Theory of Knowledge.* London: New Left.

———. 1976. *Wider den Methodenzwang: Skizze einer anarchistischen Erkenntnistheorie.* Trans. Hermann Vetter. Ed. Jürgen Habermas, Dieter Henrich, and Niklas Luhmann. Frankfurt am Main: Suhrkamp.

———. 1995. *Killing Time.* Chicago: University of Chicago Press.

Fleck, Ludwik. 1979. *The Genesis and Development of a Scientific Fact.* Trans. Fred Bradley and Thaddeus J. Trenn. Chicago: University of Chicago Press. [*Entstehung und Entwicklung einer wissenschaftlichen Tatsache: Einführung in die Lehre vom Denkstil und Denkkollektiv* (1935). Frankfurt am Main: Suhrkamp, 1980.]

———. 1983. "Zur Krise der 'Wirklichkeit'" [1929]. In *Erfahrung und Tatsache: Gesammelte Aufsätze.* Frankfurt am Main: Suhrkamp, pp. 46–58.

Foucault, Michel. 1972. *The Archaeology of Knowledge and the Disourse on Language.* Trans. A. M. Sheridan Smith. New York: Pantheon. [*L'archéologie du savoir.* Paris: Gallimard, 1969; *L'ordre du discours.* Paris: Gallimard, 1971.]

———. 1989. *The Order of Things: An Archaeology of the Human Sciences.* Trans. A. M. Sheridan. New York: Routledge. [*Les mots et les choses: Une archéologie des sciences humaines.* Paris: Gallimard, 1966.]

———. 1994. "Sur les façons d'écrire l'histoire" [1969]. In *Dits et écrits.* Paris: Gallimard, vol. 1 (1954–69), pp. 585–600.

———. 2003. *The Birth of the Clinic. An Archaeology of Medical Perception.* Trans. A. M. Sheridan. New York: Routledge. [*Naissance de la clinique. Une archéologie du regard médical.* Paris: Presses Universitaires de France, 1963.]

———. 2006. *The History of Madness.* Trans. Jonathan Murphy and Jean Khalfa. London: Routledge. [*Folie et déraison: Histoire de la folie à l' âge classique.* Paris: Plon, 1961.]

Hacking, Ian. 1983. *Representing and Intervening: Introductory Topics in the Philosophy of Natural Science.* Cambridge: Cambridge University Press.

Heidegger, Martin. 1977. "The Age of the World Picture." In *The Question Concerning Technology and Other Essays.* Trans. William Lovitt. New York: Harper and Row, pp. 115–54. ["Die Zeit des Weltbildes" (1938). In *Holzwege,* vol. 5 of *Gesamtausgabe.* Frankfurt am Main: Vittorio Klostermann, 1950, pp. 75–113.]

———. 1996. *Being and Time.* Trans. Joan Stambaugh. New York: State University of New York Press. [*Sein und Zeit.* Halle: M. Niemeyer, 1927.]

Hertz, Heinrich. 2007. *The Principles of Mechanics Presented in a New Form.* Trans. D. E. Jones and J. T. Walley. New York: Cosimo. [*Die Prinzipien der Mechanik in neuem Zusammenhange dargestellt.* Leipzig: Barth, 1894.]

Husserl, Edmund. 1970a. *The Crisis of European Sciences and Transcendental Phenomenology.* Trans. David Carr. Evanston, Ill.: Northwestern University Press. [*Die Krisis der europäischen Wissenschaften und die transzendentale Phänomenologie* (1936). *Husserliana: Gesammelte Werke,* vol. 6, ed. Walter Biemel. The Hague: Martinus Nijhoff, 1976.]

———. 1970b. "Die Krisis des europäischen Menschentums und die Philosophie" [1935]. In *Die Krisis der europäischen Wissenschaften und die transzendentale Phänomenologie. Husserliana: Gesammelte Werke,* vol. 6, ed. Walter Biemel. The Hague: Martinus Nijhoff, 1976, pp. 314–48.

———. 1978. "The Origin of Geometry." In *Edmund Husserl's "Origin of Geometry": An Introduction,* by Jacques Derrida, trans. John P. Leavey, Jr. New York: Harvester. ["Der Ursprung der Geometrie als intentionalhistorisches Problem" (1939). In *Die Krisis der europäischen Wissenschaften und die transzendentale Phänomenologie. Husserliana: Gesammelte Werke,* vol. 6, ed. Walter Biemel. The Hague: Martinus Nijhoff, 1976.]

———. 1993. "Die Naivität der Wissenschaft." In *Die Krisis der europäischen Wissenschaften und die transzendentale Phänomenologie. Husserliana: Gesammelte Werke,* vol. 29, ed. Reinhold N. Smid. Dordrecht: Kluwer Academic Publishers, 1993, pp. 27–36.

Koyré, Alexandre. 1957. *From the Closed World to the Infinite Universe.* Baltimore: Johns Hopkins University Press.

———. 1966a. "Les étapes de la cosmologie scientifique." In *Études d'histoire de la pensée scientifique.* Paris: Presses Universitaires de France.

———. 1966b. "Galilée et la révolution scientifique du XVII siècle." In *Études d'histoire de la pensée scientifique.* Paris: Presses Universitaires de France.

———. 1978. *Galileo Studies.* Trans. John Mepham. Atlantic Highlands, N.J.: Humanities. [*Études Galiléennes.* Paris: Hermann, 1939.]

Kubler, George. 1962. *The Shape of Time: Remarks on the History of Things.* New Haven, Conn.: Yale University Press.

Kuhn, Thomas. 1962. *The Structure of Scientific Revolutions.* Chicago: University of Chicago Press.

———. 1992. *The Trouble with the Historical Philosophy of Science: An Occasional Publication of the Department of the History of Science of Harvard University.* Cambridge, Mass.: Harvard University Press.

Latour, Bruno. 1993. *We Have Never Been Modern.* New York: Harvester Wheatsheaf. [*Nous n'avons jamais été modernes.* Paris: La Découverte, 1991.]

———. 1999."Circulating Reference: Sampling the Soil in the Amazon Forest." In *Pandora's Hope.* Cambridge, Mass.: Harvard University Press, pp. 24–79. ["Le 'pédofil' de Boa Vista—montage photo-philosophique." In *La clef de Berlin.* Paris: La Découverte, 1993.]

Latour, Bruno, and Steve Woolgar. 1979. *Laboratory Life: The Social Construction of Scientific Facts.* Beverley Hills, Calif.: Sage.

Mach, Ernst. 1959. *The Analysis of Sensations and the Relation of the Physical to the Psychical.* Trans. C. M. Williams. New York: Dover. [*Die Analyse der Empfindungen und das Verhältnis des Physischen zum Psychischen* (1896). 9th ed. Jena: Fischer, 1922.]

———. 1960. *The Science of Mechanics.* Trans. T. J. McCormack. Lasalle, Ill.: Open Court. [*Die Mechanik in ihrer Entwicklung, historisch-kritisch dargestellt* (1883). Darmstadt: Wissenschaftliche Buchgesellschaft, 1976 (reprint of 9th ed., Leipzig: Brockhaus, 1933).]

Neurath, Otto. 1973. "On the Foundations of the History of Optics." In *Empiricism and Sociology,* trans. Paul Foulkes and Marie Neurath. Dordrecht: D. Reidel, pp. 101–12. ["Prinzipielles zur Geschichte der Optik." *Archiv für die Geschichte der Naturwissenschaften und der Technik* 5 (1915): 371–89.]

———. 1983. "Classification of Systems of Hypotheses." In *Philosophical Papers, 1913–1946.* Trans. Robert S. Cohen. Dordrecht: D. Reidel. ["Zur Klassifikation von Hypothesensystemen (mit besonderer Berücksichtigung der Optik)." *Jahrbuch der Philosophischen Gesellschaft an der Universität zu Wien* (1915): 38–63.]

Poincaré, Henri. 1905. *La valeur de la science.* Paris: Flammarion.

———. 1952. *Science and Hypothesis.* Trans. William John Greenstreet. New York: Dover. [*La science et l'hypothèse.* Paris: Flammarion, 1902.]

———. 1958. *The Value of Science.* Trans. G. B. Halsted. New York: Dover.

Popper, Karl. 1968. *The Logic of Scientific Discovery.* London: Hutchinson. [*Logik der Forschung.* Vienna: Julius Springer, 1935.]

———. 1972. *Objective Knowledge: An Evolutionary Approach.* Oxford: Clarendon.

Rheinberger, Hans-Jörg. 1997. *Toward a History of Epistemic Things: Synthesizing Proteins in the Test Tube.* Stanford, Calif.: Stanford University Press.

Toulmin, Stephen. 1961. *Foresight and Understanding: An Enquiry into the Aims of Science.* Bloomington: Indiana University Press.

———. 1972. *Human Understanding.* Princeton, N.J.: Princeton University Press.

Wind, Edgar. 2001. *Experiment and Metaphysics.* Trans. Matthew Rampley. Oxford: European Humanities Research Centre. [*Das Experiment und die Metaphysik* (1934). Frankfurt am Main: Suhrkamp 2001.]

Index of Names

Althusser, Louis, 2, 65, 69, 73,–74
Aristotle, 46

Bachelard, Gaston, 2, 19, 20, 21–27, 29–33
Barnes, Barry, 86
Bloor, David, 86
Bohr, Niels, 29
Boutroux, Émile, 2, 10–12
Broglie, Louis de, 27
Butterfield, Herbert, 53, 61

Canguilhem, Georges, 2, 49, 65–69, 72, 75
Carnap, Rudolf, 30, 35
Cassirer, Ernst, 2, 46–49
Cavaillès, Jean, 65, 74, 75
Cohen, Bernard, 52
Collins, Harry, 86
Comte, Auguste, 1, 7

Darwin, Charles, 60
Derrida, Jacques, 2, 41, 65, 74–77
Descartes, René, 46, 53
Dijksterhuis, Eduard, 66
Dilthey, Wilhelm, 12, 14, 15, 20
Du Bois-Reymond, Emil, 2, 5–8, 10, 20
Duhem, Pierre, 5–6

Einstein, Albert, 20

Feigl, Herbert, 35
Feyerabend, Paul, 2, 61–64
Fink, Eugen, 39

Fleck, Ludwik, 2, 19–20, 27–33, 35, 39, 54, 55, 57, 61, 63, 82
Foucault, Michel, 2, 65, 69–76, 80, 85

Galilei, Galileo, 52, 53
Galison, Peter, 86
Grmek, Mirko, 86
Grossmann, Henryk, 20
Guillemin, Roger, 82

Hacking, Ian, 2, 79–83, 85, 86
Hegel, Georg Wilhelm Friedrich, 26
Heidegger, Martin, 2, 42–46, 48, 52, 69, 82
Helmholtz, Hermann von, 9
Hertz, Heinrich, 82
Hessen, Boris, 20
Hilbert, David, 52
Holmes, Frederic, 86
Husserl, Edmund, 2, 9, 39, 42, 45, 46, 48, 52, 72, 74–77
Hyppolite, Jean, 69

Kant, Immanuel, 22
Koyré, Alexandre, 2, 51–53, 54, 55
Kraft, Victor, 61
Kubler, George, 55
Kuhn, Thomas, 2, 31, 39, 52, 54–58, 59, 60, 61, 63, 68, 79–80

Latour, Bruno, 2, 79, 82–86
Leibniz, Gottfried Wilhelm, 46
Lenin, Vladímir Iljitsch, 61
Lord Acton, 36

Mach, Ernst, 1, 2, 7–10, 12, 13, 15, 16, 27, 30
Marx, Karl, 20, 70, 73, 74
Mendel, Gregor, 66
Merleau-Ponty, Maurice, 69, 74
Metzger, Hélène, 65
Meyerson, Émile, 52

Neurath, Otto, 14–17, 35
Novalis, 35

Pearson, Karl, 93
Pickering, Andrew, 86
Poincaré, Henri, 2, 10, 12–14, 15, 17, 25
Popper, Karl, 2, 35–39, 44, 45, 55, 58, 59, 61, 62

Reichenbach, Hans, 55
Rey, Abel, 21
Riezler, Kurt, 27

Sartre, Jean-Paul, 74
Schaffer, Simon, 86
Schlick, Friedrich Albert Moritz, 35, 39

Toulmin, Stephen, 2, 39, 59–61, 62, 63
Tran Duc Thao, 74

Wind, Edgar, 23
Wise, Norton, 86
Wittgenstein, Ludwig, 59
Woolgar, Steve, 82

Cultural Memory | *in the Present*

Jacob Taubes, *From Cult to Culture*, edited by Charlotte Fonrobert and Amir Engel

Peter Hitchcock, *The Long Space: Transnationalism and Postcolonial Form*

Lambert Wiesing, *Artificial Presence: Philosophical Studies in Image Theory*

Jacob Taubes, *Occidental Eschatology*

Freddie Rokem, *Philosophers and Thespians: Thinking Performance*

Roberto Esposito, *Communitas: The Origin and Destiny of Community*

Vilashini Cooppan, *Worlds Within: National Narratives and Global Connections in Postcolonial Writing*

Josef Früchtl, *The Impertinent Self: A Heroic History of Modernity*

Frank Ankersmit, Ewa Domanska, and Hans Kellner, eds., *Re-Figuring Hayden White*

Michael Rothberg, *Multidirectional Memory: Remembering the Holocaust in the Age of Decolonization*

Jean-François Lyotard, *Enthusiasm: The Kantian Critique of History*

Ernst van Alphen, Mieke Bal, and Carel Smith, eds., *The Rhetoric of Sincerity*

Stéphane Mosès, *The Angel of History: Rosenzweig, Benjamin, Scholem*

Alexandre Lefebvre, *The Image of the Law: Deleuze, Bergson, Spinoza*

Samira Haj, *Reconfiguring Islamic Tradition: Reform, Rationality, and Modernity*

Diane Perpich, *The Ethics of Emmanuel Levinas*

Marcel Detienne, *Comparing the Incomparable*

François Delaporte, *Anatomy of the Passions*

René Girard, *Mimesis and Theory: Essays on Literature and Criticism, 1959–2005*

Richard Baxstrom, *Houses in Motion: The Experience of Place and the Problem of Belief in Urban Malaysia*

Jennifer L. Culbert, *Dead Certainty: The Death Penalty and the Problem of Judgment*

Samantha Frost, *Lessons from a Materialist Thinker: Hobbesian Reflections on Ethics and Politics*

Regina Mara Schwartz, *Sacramental Poetics at the Dawn of Secularism: When God Left the World*

Gil Anidjar, *Semites: Race, Religion, Literature*

Ranjana Khanna, *Algeria Cuts: Women and Representation, 1830 to the Present*

Esther Peeren, *Intersubjectivities and Popular Culture: Bakhtin and Beyond*

Eyal Peretz, *Becoming Visionary: Brian De Palma's Cinematic Education of the Senses*

Diana Sorensen, *A Turbulent Decade Remembered: Scenes from the Latin American Sixties*

Hubert Damisch, *A Childhood Memory by Piero della Francesca*

José van Dijck, *Mediated Memories in the Digital Age*

Dana Hollander, *Exemplarity and Chosenness: Rosenzweig and Derrida on the Nation of Philosophy*

Asja Szafraniec, *Beckett, Derrida, and the Event of Literature*

Sara Guyer, *Romanticism After Auschwitz*

Alison Ross, *The Aesthetic Paths of Philosophy: Presentation in Kant, Heidegger, Lacoue-Labarthe, and Nancy*

Gerhard Richter, *Thought-Images: Frankfurt School Writers' Reflections from Damaged Life*

Bella Brodzki, *Can These Bones Live? Translation, Survival, and Cultural Memory*

Rodolphe Gasché, *The Honor of Thinking: Critique, Theory, Philosophy*

Brigitte Peucker, *The Material Image: Art and the Real in Film*

Natalie Melas, *All the Difference in the World: Postcoloniality and the Ends of Comparison*

Jonathan Culler, *The Literary in Theory*

Michael G. Levine, *The Belated Witness: Literature, Testimony, and the Question of Holocaust Survival*

Jennifer A. Jordan, *Structures of Memory: Understanding German Change in Berlin and Beyond*

Christoph Menke, *Reflections of Equality*

Marlène Zarader, *The Unthought Debt: Heidegger and the Hebraic Heritage*

Jan Assmann, *Religion and Cultural Memory: Ten Studies*

David Scott and Charles Hirschkind, *Powers of the Secular Modern: Talal Asad and His Interlocutors*

Gyanendra Pandey, *Routine Violence: Nations, Fragments, Histories*

James Siegel, *Naming the Witch*

J. M. Bernstein, *Against Voluptuous Bodies: Late Modernism and the Meaning of Painting*

Theodore W. Jennings, Jr., *Reading Derrida / Thinking Paul: On Justice*

Richard Rorty and Eduardo Mendieta, *Take Care of Freedom and Truth Will Take Care of Itself: Interviews with Richard Rorty*

Jacques Derrida, *Paper Machine*

Renaud Barbaras, *Desire and Distance: Introduction to a Phenomenology of Perception*

Jill Bennett, *Empathic Vision: Affect, Trauma, and Contemporary Art*

Ban Wang, *Illuminations from the Past: Trauma, Memory, and History in Modern China*

James Phillips, *Heidegger's Volk: Between National Socialism and Poetry*

Frank Ankersmit, *Sublime Historical Experience*

István Rév, *Retroactive Justice: Prehistory of Post-Communism*

Paola Marrati, *Genesis and Trace: Derrida Reading Husserl and Heidegger*

Krzysztof Ziarek, *The Force of Art*

Marie-José Mondzain, *Image, Icon, Economy: The Byzantine Origins of the Contemporary Imaginary*

Cecilia Sjöholm, *The Antigone Complex: Ethics and the Invention of Feminine Desire*

Jacques Derrida and Elisabeth Roudinesco, *For What Tomorrow . . . : A Dialogue*

Elisabeth Weber, *Questioning Judaism: Interviews by Elisabeth Weber*

Jacques Derrida and Catherine Malabou, *Counterpath: Traveling with Jacques Derrida*

Martin Seel, *Aesthetics of Appearing*

Nanette Salomon, *Shifting Priorities: Gender and Genre in Seventeenth-Century Dutch Painting*

Jacob Taubes, *The Political Theology of Paul*

Jean-Luc Marion, *The Crossing of the Visible*

Eric Michaud, *The Cult of Art in Nazi Germany*

Anne Freadman, *The Machinery of Talk: Charles Peirce and the Sign Hypothesis*

Stanley Cavell, *Emerson's Transcendental Etudes*

Stuart McLean, *The Event and Its Terrors: Ireland, Famine, Modernity*

Beate Rössler, ed., *Privacies: Philosophical Evaluations*

Bernard Faure, *Double Exposure: Cutting Across Buddhist and Western Discourses*

Alessia Ricciardi, *The Ends of Mourning: Psychoanalysis, Literature, Film*

Alain Badiou, *Saint Paul: The Foundation of Universalism*

Gil Anidjar, *The Jew, the Arab: A History of the Enemy*

Jonathan Culler and Kevin Lamb, eds., *Just Being Difficult? Academic Writing in the Public Arena*

Jean-Luc Nancy, *A Finite Thinking*, edited by Simon Sparks

Theodor W. Adorno, *Can One Live after Auschwitz? A Philosophical Reader*, edited by Rolf Tiedemann

Patricia Pisters, *The Matrix of Visual Culture: Working with Deleuze in Film Theory*

Andreas Huyssen, *Present Pasts: Urban Palimpsests and the Politics of Memory*

Talal Asad, *Formations of the Secular: Christianity, Islam, Modernity*

Dorothea von Mücke, *The Rise of the Fantastic Tale*

Marc Redfield, *The Politics of Aesthetics: Nationalism, Gender, Romanticism*

Emmanuel Levinas, *On Escape*

Dan Zahavi, *Husserl's Phenomenology*

Rodolphe Gasché, *The Idea of Form: Rethinking Kant's Aesthetics*

Michael Naas, *Taking on the Tradition: Jacques Derrida and the Legacies of Deconstruction*

Herlinde Pauer-Studer, ed., *Constructions of Practical Reason: Interviews on Moral and Political Philosophy*

Jean-Luc Marion, *Being Given That: Toward a Phenomenology of Givenness*

Theodor W. Adorno and Max Horkheimer, *Dialectic of Enlightenment*

Ian Balfour, *The Rhetoric of Romantic Prophecy*

Martin Stokhof, *World and Life as One: Ethics and Ontology in Wittgenstein's Early Thought*

Gianni Vattimo, *Nietzsche: An Introduction*

Jacques Derrida, *Negotiations: Interventions and Interviews, 1971–1998*, ed. Elizabeth Rottenberg

Brett Levinson, *The Ends of Literature: The Latin American "Boom" in the Neoliberal Marketplace*

Timothy J. Reiss, *Against Autonomy: Cultural Instruments, Mutualities, and the Fictive Imagination*

Hent de Vries and Samuel Weber, eds., *Religion and Media*

Niklas Luhmann, *Theories of Distinction: Re-Describing the Descriptions of Modernity*, ed. and introd. William Rasch

Johannes Fabian, *Anthropology with an Attitude: Critical Essays*

Michel Henry, *I Am the Truth: Toward a Philosophy of Christianity*

Gil Anidjar, *"Our Place in Al-Andalus": Kabbalah, Philosophy, Literature in Arab-Jewish Letters*

Hélène Cixous and Jacques Derrida, *Veils*

F. R. Ankersmit, *Historical Representation*

F. R. Ankersmit, *Political Representation*

Elissa Marder, *Dead Time: Temporal Disorders in the Wake of Modernity (Baudelaire and Flaubert)*

Reinhart Koselleck, *The Practice of Conceptual History: Timing History, Spacing Concepts*

Niklas Luhmann, *The Reality of the Mass Media*

Hubert Damisch, *A Theory of /Cloud/: Toward a History of Painting*

Jean-Luc Nancy, *The Speculative Remark: (One of Hegel's bon mots)*

Jean-François Lyotard, *Soundproof Room: Malraux's Anti-Aesthetics*

Jan Patočka, *Plato and Europe*

Hubert Damisch, *Skyline: The Narcissistic City*

Isabel Hoving, *In Praise of New Travelers: Reading Caribbean Migrant Women Writers*

Richard Rand, ed., *Futures: Of Jacques Derrida*

William Rasch, *Niklas Luhmann's Modernity: The Paradoxes of Differentiation*

Jacques Derrida and Anne Dufourmantelle, *Of Hospitality*

Jean-François Lyotard, *The Confession of Augustine*

Kaja Silverman, *World Spectators*

Samuel Weber, *Institution and Interpretation: Expanded Edition*

Jeffrey S. Librett, *The Rhetoric of Cultural Dialogue: Jews and Germans in the Epoch of Emancipation*

Ulrich Baer, *Remnants of Song: Trauma and the Experience of Modernity in Charles Baudelaire and Paul Celan*

Samuel C. Wheeler III, *Deconstruction as Analytic Philosophy*

David S. Ferris, *Silent Urns: Romanticism, Hellenism, Modernity*

Rodolphe Gasché, *Of Minimal Things: Studies on the Notion of Relation*

Sarah Winter, *Freud and the Institution of Psychoanalytic Knowledge*

Samuel Weber, *The Legend of Freud: Expanded Edition*

Aris Fioretos, ed., *The Solid Letter: Readings of Friedrich Hölderlin*

J. Hillis Miller / Manuel Asensi, *Black Holes / J. Hillis Miller; or, Boustrophedonic Reading*

Miryam Sas, *Fault Lines: Cultural Memory and Japanese Surrealism*

Peter Schwenger, *Fantasm and Fiction: On Textual Envisioning*

Didier Maleuvre, *Museum Memories: History, Technology, Art*

Jacques Derrida, *Monolingualism of the Other; or, The Prosthesis of Origin*

Andrew Baruch Wachtel, *Making a Nation, Breaking a Nation: Literature and Cultural Politics in Yugoslavia*

Niklas Luhmann, *Love as Passion: The Codification of Intimacy*

Mieke Bal, ed., *The Practice of Cultural Analysis: Exposing Interdisciplinary Interpretation*

Jacques Derrida and Gianni Vattimo, eds., *Religion*

The authorized representative in the EU for product safety and compliance is:
Mare Nostrum Group
B.V Doelen 72
4831 GR Breda
The Netherlands

www.ingramcontent.com/pod-product-compliance
Lightning Source LLC
Chambersburg PA
CBHW032301150426
43195CB00008BA/539